KID FRIENDLY COMPUTATION

MULTIPLICATION & DIVISION

SARAH MORGAN MAJOR

Zephyr Press

Chicago

Library of Congress Cataloging-in-Publication Data

Major, Sarah Morgan, 1953-
    Multiplication and division / Sarah Morgan Major.
        p. cm. — (Kid-friendly computation)
    ISBN 1-56976-196-5
1. Multiplication—Study and teaching (Elementary)—Audio-visual aids. 2.
Division—Study and teaching (Elementary)—Audio-visual aids. I. Title.
QA115.M23 2005
    372.7'2—dc22   2005002764

Cover and interior design: Rattray Design
Illustrations: Sarah Morgan Major

© 2005 by Sarah Morgan Major
All rights reserved
Published by Zephyr Press
An imprint of Chicago Review Press, Incorporated
814 North Franklin Street
Chicago, Illinois 60610
ISBN 1-56976-196-5
Printed in the United States of America

Contents

Introduction

One particular train of thought has been haunting me, widening and deepening throughout the years in which I have been working with children who struggle in regular classrooms. That train of thought involves the elements that seem to enable and inspire struggling children to learn and remember. The ideas to which I refer involve the power of the visual; the magic way in which the body remembers motion; the skill and mastery that come from making rapid, meaningful connections; and the bent of the human mind toward recognizing patterns and organizing individual parts of the world into groups of elements that have similar or overlapping characteristics.

In my years of teaching, I have witnessed students who once consistently failed begin to make progress—while investing a fraction of the time they'd previously spent studying. How does this happen? What change has caused, or allowed, these students to excel in areas that once seemed beyond them? The more I've studied the phenomenon, the more I've lost my loyalty to the way in which "things have always been done." For me, the impetus for change in my teaching methods has been a desire to help students break down the barriers to learning; to help them achieve their highest potential.

Our most vivid memories are ones in which every detail is recorded, to the point that we can almost relive the event. The ideas that we remember best are usually accompanied by a particular emotion, but equally powerful are the pictures of events that our brain has recorded. We say, "I can see exactly where I was when that happened. I remember the color you were wearing—the look on your face." The power of the visual to serve as context for recall is underestimated on a regular basis in our traditional teaching methods. For the sake of speed, we want students to just "study and remember" abstract and unrelated facts, and we don't take the time to embed these isolated facts into a picture, a whole, a pattern that would make learning possible and memory permanent. In teaching advanced vocabulary terms to middle schoolers, I found that by embedding both the spelling and the meaning of a word into a picture, students who had previously failed vocabulary and spelling assessments not only learned quickly, but succeeded in recalling the material for their tests.

We take for granted the incredible ability of our cerebellums to store intricate information about motion. Witness the skill of Olympic gymnasts on the exercise mat. What they are able to learn and then perform, seemingly unconsciously, is astonishing. In truth, however, every human possesses the ability to store memories of complex series of motions and to reproduce those motions flawlessly. When a pianist memorizes an intricate piece for a recital, the replaying for the audience involves mostly the outflow of the motions that have been learned by the body. The cerebrum is not heavily involved at that point. I am afraid that most of the time, we associate physical giftedness with simply physical strength, endurance, and innate coordination, while failing to recognize the untapped potential of directly tying motion to concepts learned.

Sometimes, the mind-body connection involves directly reproducing the shape of the concept. For example, in my reading program, when students learn the alphabet, I have them reproduce the letter shape with their bodies. These motions are immensely powerful for struggling learners who have difficulty recalling sounds and tying them to letter symbols. When visual learners try to recall a concept that they have learned, they frequently unconsciously reproduce the motions they made while learning the concept. In those cases, it is the motion associated with the concept that serves as a vehicle for recalling the abstract concept.

Our world is full of patterns. I can see in my mind's eye the pattern found inside a nautilus shell, as well as the pattern found in the array of sunflower seeds at the heart of the bloom—there truly is no limit to the patterns found in our world. We are drawn to these patterns; they delight and fascinate us. When we are presented with random objects, our tendency is to organize them into a pattern—say, from short to tall, or from small to large. The Kid-Friendly Computation method is all about learning by tapping into the incredible power of patterns that are found when numbers are arranged in specific ways. Using this method, students actually learn and recall facts because of where they are found within the pattern. When I ask them how they recalled a specific fact, more often than not their answer has something to do with location within a global pattern.

Just as we are drawn to patterns, we also tend to want to organize everything in our world according to likeness. In the kitchen, food tends to be grouped together; plates and cups are in close proximity; cleaning supplies are stored together; and so forth. In our dressers, we separate socks from sweaters. If we utilize this penchant for organizing as we teach concepts to our students, we will have harnessed another powerful tool. It is far easier to teach like concepts together when the likeness between the individual concepts stands out in bold relief than it is to teach random facts together. For example, my second graders were learning 10 spelling words a week. When I began to group together words that contain a similar element—such as a particular sound spelling—they not only learned rapidly, but were also able to double the number of words they could spell. One sound spelling is /oi/. Consider one of the shortest words that contains that—"oil." Once students are taught that small word, a number of larger words that contain the same sound spelling—soil, boil, spoil, moist, point, coins, and so forth—may be easily taught. Students begin to rely upon patterns and the organization of words according to similar elements, and they rely heavily on their sense of sound to figure out the remaining letters in the words.

The *Kid-Friendly Computation* series relies on all of these elements. The contents of this particular book on multiplication and division focus on seeking patterns in five-column charts. Traditionally, numbers are arrayed in rows of ten numbers. This is fine for some purposes, but the patterns found in five-column charts are truly magical.

You will also find that some of the handouts in this book contain duplicates of problems. This is intentional; when students encounter the same combination of numbers repeatedly, they begin to recognize "familiar faces," their "Aha!" light goes on, and what results is fluency and confidence. This fluency is the basis for learning and automaticity in their math. Frequent repetition of problems allows students to become "experts" with the facts. Whatever we practice, we master, and because basic computation is the foundation for more advanced mathematics, gaining fluency is especially important at this stage.

These concepts are quite different from traditional approaches to teaching multiplication and division. I want to challenge and encourage you to delve into this new approach and see where the ride takes you and your students! And I would encourage all teachers to become passionate about searching out ways to use these powerful tools in teaching any material, regardless of the subject or grade level. At first, it may seem like a lot of trouble, and it may come with some difficulty, but the more you persist in this line of thinking, the more automatic this approach will become and the more easily the ideas will flow. Before long, this way of teaching will become second nature. Get ready for the results—they will amaze you!

1

Magic Fives

Goals for This Chapter

1. Construct a five-column chart

2. Learn number placement within a blank five-column chart

3. Discover number patterns within a five-column chart

4. Become fluent in locating numbers within a five-column chart

To introduce students to numbers arrayed in a five-column chart is to invite them to experience the order and relationships between numbers in a grid. To extend the use of the five-column chart into the realm of multiplication and division is to introduce children to the wonder of myriad patterns that will delight young minds and ease the learning of dreaded multiplication and division facts. Bid a hasty good-bye to traditional methods of teaching multiplication and division, throw out those flash cards to be memorized, and welcome a new method that entices students to actively participate in creating meaning in what they learn instead of reluctantly memorizing seemingly abstract products and quotients.

In order to attain this result, it would not be best practice to stand before the class one morning and announce, "Students, today we are going to start the study of multiplication and division." Instead, approach the subject by doing a simple exercise together and by playing some disarmingly simple games. The primary tool students will need is their ability to notice, or an active "noticer." They can place their "memorizer" into cold storage for now; we will not need it until chapter 7. Students are often passive bystanders in the classroom. They listen (or don't listen) to what the teacher tells them, do (or don't do) the homework they are given, and memorize (or don't memorize) the facts they are asked to "learn." But if students have not been proactive in their search for knowledge and there is no understanding of what is being learned, their long-term acquisition of knowledge will be minimal. As a teacher, you want to make students' learning as rich, deep, and meaningful as possible. You want students to be searchers and delvers who uncover meaning because they want to.

Introduce

Ask your students to think of all the objects that come in groups of five that they can possibly identify. Write their ideas on the board. Famous "fives" include fingers on their hands, toes on their feet, points on a star, minutes between numbers on a clock, pennies in a nickel, and nickels in a quarter. Let the question hang for a couple of days, and leave the list on the board to add to as children investigate the question outside of class.

Investigate

Give each student a strip of grid paper containing 20 half-inch boxes, five boxes wide by four boxes deep. If you prefer to photocopy grid sheets, use blackline master S1.1 (page 85). You may also show them blackline master S1.2 (page 85) on an overhead projector. Working together as a class, fill in the numbers on the grids as shown below.

1	2	3	4	5
6	7	8	9	10
11	12	13	14	15
16	17	18	19	20

Empty fives grid.

Ask students what they notice about the numbers in the grid they just filled in. Some patterns their "noticers" might pick up on include the following:

- In the first column, the last digit of each number forms this pattern: 1, 6, 1, 6.
- In the second column, the last digit of each number forms this pattern: 2, 7, 2, 7.
- In the third column, the last digit of each number forms this pattern: 3, 8, 3, 8.
- In the fourth column, the last digit of each number forms this pattern: 4, 9, 4, 9.
- In the fifth column, the last digit of each number forms this pattern: 5, 0, 5, 0.

Looking at the tens place, the pattern is: 0, 0, 1, 1 in columns one through four, while in the fifth column the pattern is 0, 1, 1, 2.

You can be fanciful in your exploration of these patterns. For example, ask students:

Do you think that maybe the number one has fallen straight down and has become curved at the bottom, forming the number six?

Do you see similarity in the numbers two and seven? Each number contains a slant from top right to bottom left. The number seven is almost like a number 2 without a foot!

Do you see that, in the third column, the number three looks like a number eight with one side cut off?

Do you see that in the "four, nine" pattern the four and the nine are both top-heavy, and that each has a line extending down to complete the number?

And that in the last column, the numbers zero and five both have rounded parts?

Here are other ways to explore the patterns found in this grid:

✐ What happens when you move from one number on the grid to the number just below it? (The number is increased by five.) Experiment to see if this pattern holds true for every number on the grid.

✐ What happens if you move directly up from any number? (The number is decreased by five.)

✐ What happens if you start at the number one and move diagonally to the right? (You land on the number seven, which is six more than the number you were on.) Does this hold true for any number on the chart?

✐ What happens if you start on five and move diagonally to the left? (You land on the number nine, which is four more than the number you were on.) Does this hold true for any number on the chart?

✐ What if you start at one number and skip straight down two rows? (The number is increased by 10.) Test this pattern for consistency.

✐ What if you skip straight up two rows? (The number is decreased by 10.) Again, test this pattern for consistency.

Ask the children if they think these "rules" they've discovered would hold true if they made a five-column chart to 40. How about to 80 or 100? If they are interested in testing these rules, let them work on a five-column chart that goes to a large number (determined by them).

Internalize

Discover

Using a five-column chart with only three rows (blackline master S1.3, page 86), review the placement of the numbers in the grid. While the children are looking at their grids, say the following:

1	2	3	4	5
6	7	8	9	10
11	12	13	14	15

✐ Find the center number (8).
✐ Find the numbers in the left corners (1, 11).

✐ Find the numbers in the right corners (5, 15).

✐ Name the numbers above and below the "magic center number" 8 (3, 13).

✐ Name the numbers above and below the number 6 on the grid (1, 11).

✐ Name the number found on the grid between the numbers 4 and 14 (9).

✐ Name the numbers found at the beginning of each row (1, 6, 11).

✐ Name the numbers found at the end of each row (5, 10, 15).

✐ Name the numbers found in the center of each row (3, 8, 13).

✐ Name the numbers found in the column next to the last column (4, 9, 14).

✐ Name the numbers found in the column next to the first column (2, 7, 12).

Practice

Give each student an empty grid (blackline master S1.1, page 85) and a pencil. Call out the numbers 1 through 15 in random order, and have students write the numbers in the correct boxes on their grids. Students may then practice this activity working independently, in pairs, in small groups, or in math centers. Let one child be the "caller" who calls out numbers at random. This is easier when the caller is given a set of number cards from 1 to 15 that are not in correct order. The caller calls a number, places the number card at the back of the stack, and calls out the next number, until all numbers have been called. Have the caller shuffle the cards and pass them to another student. Repeat the activity.

Challenge

A variation of this practice activity involves empty grids and colored plastic chips or other markers of some kind. As one student calls the numbers at random, the partner or group places a chip in the box in which it would be found.

To make the activity more challenging, have the caller identify the number in question as:

✐ the number over the number 7;

✐ the number under 8;

✐ the number over 9;

✐ the number under 1;

✐ the number over 10, and so forth.

As children practice these exercises, they gain mastery of knowing where the numbers are located in relation to one another on the five-column chart.

Assess

Give students an empty grid and a pencil and have them put their names on their papers. Call numbers from 1 to 15 at random and have students write them in the correct boxes in their grids. Do a quick visual check to see who placed the numbers correctly.

Alternatively, have students do the same activity in small groups so that you can watch them as they work. Notice which children hesitate or have to work hard to figure out where the numbers go. You do not want any child to accomplish this task by swiftly "counting up" to the correct number on the grid!

When you have determined that the class has mastered the five-column chart, move on to chapter 2.

Mirror Images

Goals for This Chapter

1. Gain an understanding of the meaning of multiplication

2. Gain an understanding of the meaning of division

3. Practice multiplication skills using real materials

4. Practice division skills using real materials

Each had 7 . . . they had 21 *in all!*

From the approach of identifying patterns, multiplication and division are taken up together, as they are mirror images of each other. Consider the numbers 3, 5, and 15. The numbers are multiplied by "walking to the right": "3 times 5 is 15." Making an about-face, the numbers are divided by "walking to the left": "15 divided by 5 is 3." Four related multiplication or division scenarios may be created from these, or any other trio of related numbers: 5 times 3 is 15; 3 times 5 is 15; 15 divided by 5 is 3; and 15 divided by 3 is 5. To put these problems into child-friendly terms, tangible items should discussed: "Five groups of 3 cookies is a total of 15 cookies," or, "15 cookies shared equally among 3 children means 5 cookies for each child." The action in multiplication involves generating a large number from two or more smaller numbers, while the action of division results from a large number being broken down into smaller parts.

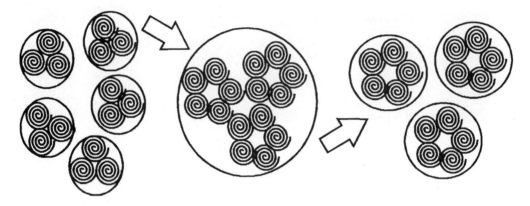

5 groups of 3 rolls   is the same as 1 group of 15 rolls   and as 3 groups of 5 rolls.

Instead of simply giving students tables to memorize, give them hands-on experience with moving real objects around so they will gain a solid understanding of what is really happening when they multiply and divide using abstract symbols. By practicing these functions with real materials, children learn naturally why they might use multiplication and division skills throughout their lives.

There are some vital differences between this constructivist-learning approach to multiplication and division and other, more traditional methods. The biggest difference is that students do not rely primarily on simple memorization of abstract facts. In addition, children do not just learn a formula that they might or might not know how to apply in other situations. By expanding teaching methods to include techniques that appeal to highly visual children, teachers allow those gifted children to succeed in an arena that has traditionally been the domain of more sequential, mathematical minds. A wonderful by-product of this method is that students enjoy the process far more than they would simply memorizing a times table. In order for *all* children to learn, use, and remember facts, teachers must carve out the time needed to do it right.

Introduce Multiplication

Materials Needed:

5 colored bowls
1 tray
15 small counting objects, such as pebbles or large dry beans
OR
Overhead projector
Blank transparency
Dry-erase marker
15 colored plastic chips

In this activity, you may use an overhead projector, a transparency, a dry-erase marker with which to draw circles on the transparency, and plastic chips. Or, you may gather the class around you on the floor and use the more friendly materials of colored bowls, a tray, and your choice of counting items. (I love using smooth, flat pebbles I've gathered from the beaches of northern Michigan—their varied sizes, textures, colors, and shapes are delightful. I have also used little plastic bears and colored plastic chips.) The tray is used to hold a pile of counting items, or the dividend, the number of bowls used represents the divisor, and the number of items in each bowl represents the quotient. Either way you choose to conduct this activity is fine. When you are ready, set the stage:

Mrs. Twig loved to bake cookies. In fact, she baked cookies almost every single day. She also had a lot of children, and they loved to eat those cookies. When they were outside playing, no matter how far away they were from the house, they could smell that lovely cinnamon smell, or that delicious chocolate smell, and they would come running! This particular day, Jane, Mrs. Twig's oldest child, was near the back door. She took a big sniff and rushed into the kitchen, slamming the screen door behind her and yelling, "Mama! Mama! May I have some cookies? I'm staaarving!" Mrs. Twig smiled and gave Jane a bowl containing three cookies that were still warm from the oven. (Place three counting items in a bowl and show it to the students. Say, "Here's a three.")

Jane had just taken a huge bite of a cookie when her brother Tommy stomped through the back door, sniffing the air quite loudly. "I smell cookies!" he shouted. "Mommy, Mommy, may I have some?" Mrs. Twig smiled, and because she had given Jane three cookies, and because she's always fair, she handed Tommy a bowl with three cookies in it as well. (Place three counting items in another bowl and show it to the students. Say, "Here's another three.")

Tommy had just gulped down his first cookie, almost in one bite, when in raced Bob and Bill. "Me too! Me too!" they shouted. Mrs. Twig smiled and gave the boys each a bowl with three cookies in it. (Put three chips in each of two more bowls and show them to the students. Say, "Here are two more threes." Put these bowls next to the first two bowls.)

Just when Jane was eating her last bite, her sister Jill crept into the kitchen, looked around, and whispered as she tugged on Mrs. Twig's skirt, "Mommy, can I have some, too?" Mrs. Twig smiled at Jill as she gave her a bowl with three cookies in it. (Put three chips into the fifth bowl, show it to the students, and say, "Here is another three.")

When the children had finished eating, they said thanks, of course, and ran off to play—except Jill, who stayed behind. "Mommy," she said, "We ate a lot of cookies!"

"Yes, Jill, five times I served a bowl of three cookies. That makes fifteen cookies altogether."

(Draw attention to the bowls and tell students that they represent these symbols: $5 \times 3 = 15$. Reiterate, "Five times I served three cookies. Altogether I served fifteen cookies.")

Investigate Multiplication

Give students problems to practice making models for. Keep in mind that some students tend to do better working independently and that some do better in small groups. The primary goal of this activity is to give each student an opportunity to do some hands-on investigation. You will provide them with the problems below, but the children may choose the way in which they show and solve the problem depending on their own learning style.

✐ **Visual/spatial** children might want to draw sketches of the action of each problem, or they might want to cut out shapes to glue into various groupings representing the problems.

✐ **Verbal/linguistic** children might enjoy writing short scenarios for each problem.

✐ **Kinesthetic** children might choose to act out the problems using tape on the floor and fellow students. The musical/rhythmic bodies might enjoy working with kinesthetic children if they are able to contribute some form of choreography to the game.

✐ **Intrapersonal** children might choose to work out the problems using the bowls and counting items that you used in the previous activity.

✐ **Interpersonal** children might want to practice teaching the lesson to their small group, with each student serving as "teacher" for one problem.

✐ Keep an eye on those **logical/mathematical** students and see what form their fancy dictates.

Here are some practice problems:

2 × 6

8 × 3

5 × 4

3 × 4

9 × 2

4 × 7

1 × 6

Introduce Division

Gather your class together once more. Have on hand 15 counting items and five bowls. Tell students the following story about division:

Do you remember Mrs. Twig? Do you remember that she loved to make cookies for Jane, Tommy, Bob, Bill, and Jill? Well, the children had been away at camp, and they were coming home the very next day! Mrs. Twig had really missed the children, and she couldn't wait to see them again. She was sure they would be hungry for her cookies after being gone a whole week, so she decided to make a big batch of cookies for their snack.

Mrs. Twig set to work, and before too long, the cookies were finished. They smelled so good! Mrs. Twig still had time to spare, so she decorated little bags, one for each child, and put their names on them. She wanted to give each child the same number of cookies, but she wasn't sure how many that would be this time. She thought for a moment, then said, "I'll count all the cookies I baked and see how many I have." So she did. "One, two, three," she counted, all the way to fifteen! "Fifteen cookies!" she said to herself. She started to put the cookies into each bag, "One for Jane, one for Tommy, one for Bob," and she kept right on, until all the cookies were gone. (Model this action using five bowls and your counters.)

When Mrs. Twig had put all the cookies in the bags, she looked into Jane's bag and saw three cookies. She peeked into the other bags, and sure enough, they also had three cookies each.

"Whew!" said Mrs. Twig. "That worked out just right! I had 15 cookies to share among five children, and they each got three."

Tip for Visual/Kinesthetic Connections

Mention that the division sign is like a hand separating (dividing) the two dots, and the multiplication sign looks like two crossed arms that are holding all the items together. Use your hands to act out these motions when you first present multiplication and division problems. The arms crossed over the chest become the "X" of the times symbol when you ask, "How many did she serve in all?" The horizontal slicing motion would accompany the story of dividing the 15 cookies into 5 bags. The division sign is like a hand cutting through a pile and making it into smaller piles, while the crossed arms gather small groups of items into one big group.

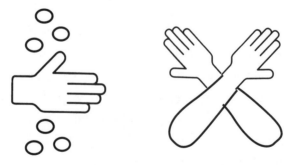

Investigate Division

Give students division problems to solve, and allow them to use real materials to work them out. Use the following sample problems, if desired:

$8 \div 2$
$12 \div 3$
$15 \div 5$
$9 \div 3$
$10 \div 2$
$6 \div 3$
$4 \div 2$

At this point, it's a good idea to show students the two ways in which a division problem may be set up. We have discussed the "dash and two dots" symbol and have seen its visual. But sometimes students will see this figure:

$$3 \overline{\smash{)}15}$$

and will need something concrete to connect with or tie to the symbol. Explain to students that the shape represents a house with a door. Inside the house is the big number. That big number could be anything they want to make it. (It could be the number of cookies that Mrs. Twig baked for her children to eat when they got home from camp. It could be a bag of candy that three children are told to split up evenly among themselves. It might be the number of dollars that will be divided among three children and given as their weekly allowance.) The number standing at the door represents the number of children knocking on the door to ask for their fair share.

When the 15 items that are "in the house" (those items may be cookies, marbles, candy, money, or some other group of objects) are divided among the three children who are "standing at the door," the number of items that each child receives is written "on top of the house":

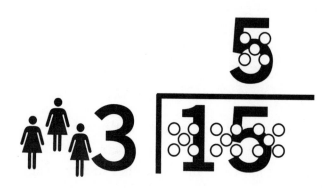

Internalize Multiplication and Division

Discover

Lead students to discern when they would use each function. When is division appropriate? When is multiplication appropriate? Have the class practice the following word problems as a group. Encourage discussion regarding the process. In this activity, do not be overly concerned with students' arriving at the right answer; the goal here is to ensure that students understand which tool (multiplication or division) is required to solve the problem.

Practice

Read a problem, then open a discussion about how that problem should be solved. In time, students will become skilled at knowing when to use each function. (Some students might indicate this by using the slicing "division" motion or the crossed-arm "multiplication" motion.)

Sample Word Problems

- 5 children have 2 doughnuts each. How many doughnuts are there altogether? (Multiplication)
- At the family reunion, 5 dogs had 2 toys each to play with. By the end of the day, all the toys were lost. How many toys did the family have to find before they left? (Multiplication)
- Under the tree on Christmas morning are 20 presents. Each person in the family has the same number of presents. If there are 5 people in the family, how many presents does each person have? (Division)
- Mrs. Brown made a pan of brownies, which she cut into 30 squares. She wants to share the brownies equally among 10 people. How many squares should she give each person? (Division)
- 6 boys are making wooden cars. They need to buy wheels for their cars. Of course, they each need 4 wheels. How many wheels do they need to buy altogether? (Multiplication)
- 8 children are making treat bags to give to sick children at the hospital. Each child makes three treat bags. How many sick children can they give treat bags to? (Multiplication)
- Mrs. Twig made 20 cookies and shared them equally among her five children. How many cookies did each child receive? (Division)

Challenge—Number Trios

Take some time to explore the idea of number trios, such as the trio of 3, 5, and 15 that has been discussed. Use the flash cards provided in this book (blackline master S2.1, pages 87–97). Displaying one flash card at a time, have

students use counting items and bowls to "work out" the problem depicted on the flash card. For example, display the flash card that shows 15 divided by 3. Ask students to use the bowls and counting chips to show you that 3 groups of 5 equals 15. Next, ask them to show you that 15 divided into 3 groups is 5. Display the same flash card again, and ask students what would happen if the 5 stood at the door instead of the 3. Would either number change? What would the number on top of the house be?

Continue to explore the concept of number trios for a while, using other trios. You want students to understand the close relationship between multiplication and division, as well as the fact that some numbers go together in threes. Other trios you might use are 3, 4, and 12; 2, 3, and 6; 3, 3, and 9; and 4, 5, and 20.

Word Problem Practice

Make up word problems and give students real materials to use in solving the problems. Have students write down their answers. In order to arrive at an answer, they will have to determine which process to use, multiplication or division.

When you are comfortable that students have mastered the concepts of multiplication and division, assess what they have learned; determine what, if anything, has not yet been understood; and allow students more practice in areas that are causing difficulty, using real materials. Remember, the goal is to ensure that every child masters these concepts.

Assess

Use the assessment provided (T2.2, page 174) to assess students' mastery of the concepts of multiplication and division. Give each student a copy of the assessment form, along with counting items, bowls, and drawing materials, if desired. The assessment is meant to evaluate students' understanding of when to use each function; do not assess students' ability to arrive at the correct answer.

Score the assessments. Record students' scores in the column labeled Concepts of M/D (multiplication and division) on the Concept Mastery Tracking Chart (T2.3, page 175).

Master Maps

> ## *Goals for This Chapter*
>
> 1. **Construct a master map of multiplication and division facts**
>
> 2. **Employ a cooperative, student-driven method of working**
>
> 3. **Engage students in the process through storytelling**
>
> 4. **Deepen students' understanding of facts to be learned by providing context**

Rather than providing students a preassembled facts chart, deepen their interest, involvement, and understanding by having them construct their own master maps. It is always good practice to let children work out problems instead of handing them something they had no part in constructing and then telling them to "learn it." By making their own maps, students are active participants in constructing meaning for their own learning.

Have students pair up and work as partners to foster a spirit of collaboration in the classroom. Each child will make his or her own chart, but having a partner facilitates their sharing of their discoveries.

Introduce

For this activity you will need the same materials as before: tray, bowls, and counters. You will also need one laminated copy of the completed Multi-Div chart (blackline master S3.2, page 99) and enough copies of T3.1 (page 176) for you and students to use. Post your copy of T3.1 where all the children can easily see it (or create a transparency from it to display on an overhead projector), then begin the story:

> Months passed, and Mrs. Twig kept making her cookies. She noticed after some time that she was doing a lot of counting and a lot of saying, "One for you, one for you, and one for you." Sometimes she noticed that she was figuring the same things out over and over again. She just couldn't seem to remember how many cookies to make if she wanted to feed several children at once.
>
> Something else was happening, too. Mrs. Twig's children were bringing more and more of their friends home to sample their mother's treats. Those good-smelling cookies were attracting all sorts of children—big kids, little ones, tall, short, loud, funny, sweet, mean—ALL kinds! The children always got cookies at Mrs. Twig's house. One afternoon, Mrs. Twig counted 12 kids around her kitchen table at one time. "I've got to do something!" Mrs. Twig thought. She decided to make some charts to help her remember the numbers she needed as she gave cookies to all those kids. This way, she wouldn't have to count as often. These are the charts that Mrs. Twig came up with. (Show the students T3.1.)
>
> First she wrote a number in the top left box of each small chart. The name of that box is the "one box." Starting with the number 1, she wrote

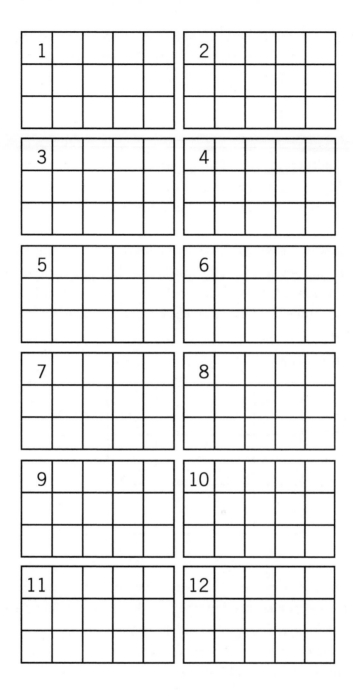

numbers in each of the one boxes of the small charts, from 1 to 12. When all the small charts were placed together, they made one master map!

Mrs. Twig planned to use the master map to figure out how many cookies she would need in order to feed all the children that she had in her kitchen at one time. On days when only her five children were there, she would look in the little chart that had a five in the one box. (Find that chart together on T3.1.)

Investigate

1. Call the students' attention to the fives chart.
2. Say, "If Mrs. Twig wanted to give the five children one cookie each, she would look in the one box. That box would tell her how many cookies she needed in all. If her children were home for cookie time and she wanted to give them two cookies each, which box would Mrs. Twig look in to see how many cookies she needed in all?" (She'd look in the two box.)
3. Ask the children to help you figure out how many cookies Mrs. Twig would need if she wanted to give five children two cookies each. (Write the number 10 in the two box.)

5	10			

4. Ask the children where to look if Mrs. Twig wanted to give each of her five children three cookies that day (the three box). How many cookies would she need that day? (She would need 15 cookies.)
5. If the children notice that they are counting up by five as they fill in numbers on the chart, let them fill in their own fives chart on their master map. First, they will need to number the charts on the master map by writing the numbers 1 to 12 in the one boxes, as you did when you told the story.
6. Before you end the lesson, have students practice identifying the meaning of the numbers on the chart. The one box indicates the number of children in Mrs. Twig's kitchen. The other boxes indicate how many total cookies are needed, depending on how many cookies Mrs. Twig wants to give each child. For example, the eight box tells you that Mrs. Twig needs 40 cookies in order to give 5 children 8 cookies each. The four box tells you that she will need 20 cookies in order to give 5 children 4 cookies each.

Here's another way to look at the chart:

Children Cookies/box	Children Cookies	Children Cookies	Children Cookies	Children Cookies
5×1 5	5×2 10	5×3 15	5×4 20	5×5 25
5×6 30	5×7 35	5×8 40	5×9 45	5×10 50
5×11 55	5×12 60	5×13 65	5×14 70	5×15 75

In all (left) ... In all (right)

Internalize

Discover

Focus next on the tens chart. Students will already have a 10 written for them in the one box. If necessary, model filling in the tens table for them, with students supplying totals for each box: "There are 10 children in the kitchen, and (point to the two box) it is a two-cookie day. How many cookies will Mrs. Twig need altogether in order to give 10 children two cookies each?" (Write the number 20 in the two box.) Have the children work in pairs to complete their tens charts and instruct them to check their work against your own filled-in chart. It will be interesting to note whether or not the children notice that the numbers in this chart are the box numbers followed by a zero.

Children Cookies/box				
10×1 10	20	30	40	50
60	70	80	90	100
110	120	130	140	150

In all

Practice

Ask the students to break into pairs to work on the even-numbered charts (2, 4, 6, 8, and 12). Each child should complete his or her own chart, but allow the pairs to discuss their work and solve the problems together. Have students start with the twos chart and proceed from there. When they have completed this project, have them check their work against your filled-in master map. Make special note of comments the children make as they work. It is important for you to hear what they are thinking, what they don't understand, what they notice about the numbers and the charts, and how they arrive at each answer.

Challenge

Have the children work with a partner to complete the odd-numbered charts on the master map. If necessary, model the threes chart for them, filling in this chart together as students figure out the correct number to put in each box.

When they have completed their threes, sevens, nines, and elevens charts, have students check their master maps against your completed master map.

Assess

The primary skill to assess at this time is accuracy in filling in master maps. It is essential that the work be neatly done so that each child can use his or her master map as a reference tool during the upcoming lessons. Copies of blackline master S3.1 (page 98) may be used for student practice prior to assessment. A completed master map appears on the next page. When you have determined that all master maps are accurate and legible, you are ready to move on to the next lesson.

Master Map for Multiplication and Division

1	2	3	4	5
6	7	8	9	10
11	12	13	14	15

2	4	6	8	10
12	14	16	18	20
22	24	26	28	30

3	6	9	12	15
18	21	24	27	30
33	36	39	42	45

4	8	12	16	20
24	28	32	36	40
44	48	52	56	60

5	10	15	20	25
30	35	40	45	50
55	60	65	70	75

6	12	18	24	30
36	42	48	54	60
66	72	78	84	90

7	14	21	28	35
42	49	56	63	70
77	84	91	98	105

8	16	24	32	40
48	56	64	72	80
88	96	104	112	120

9	18	27	36	45
54	63	72	81	90
99	108	117	126	135

10	20	30	40	50
60	70	80	90	100
110	120	130	140	150

11	22	33	44	55
66	77	88	99	110
121	132	143	154	165

12	24	36	48	60
72	84	96	108	120
132	144	156	168	180

Many Addresses

Goals for This Chapter
1. Develop competency in locating answers on the master map
2. Practice making multiplication problems from numbers in boxes
3. Practice making division problems from numbers in boxes

Before moving into pattern discovery, spend some time letting students become familiar with the way in which the master map works. The more you expose students to the contents of the map through games and activities, the more quickly they will be able to navigate within its boundaries. The individual boxes on the map almost resemble addresses on a city map, which are located using numerical coordinates. Your goal is to guide students to develop such familiarity with the map that when you present them with a word problem, they will be able to move rapidly to the particular box that contains the answer to the problem. This fluency makes the final internalizing of each times table a far easier task.

Introduce

For an introductory warm-up to this lesson, gather the students around you. Have on hand a large copy of the master map for multiplication and division as you tell the students a story:

> Mrs. Twig was very proud of her new map of numbers, and she posted it on her kitchen wall. At night, however, after the children were in bed, she would take the map off the wall and explore it as she sipped her cup of tea. Because she played with the map so often, Mrs. Twig became very quick at locating numbers for cookies, and she even started to remember some of the numbers without having to look at her beloved map.
>
> Mrs. Twig's children noticed that she wasn't consulting her master map as often when she gave them their cookies, so they asked her about it. "Mom, why didn't you look at your map when you got our cookies this time?" Mrs. Twig stopped in surprise.
>
> "Why, I didn't look, did I? I just remembered this time." Her children were so interested in what was happening with their mother that they started coming in at odd times and asking her questions just to see what she would do.
>
> "Mom, 6 kids are at your table and it is a 7-cookie day. How many cookies do you need?"
>
> "Mom, 9 kids are at your table and it is a 5-cookie day. How many cookies do you need?"
>
> "Mom, 11 kids; 13 cookies. How many?"
>
> "Mom, 7 kids; 8 cookies!"

Sometimes Mrs. Twig would point to a box on her master map, but more and more often she could tell them the answers without looking at her map at all.

Investigate

Display your large master map and ask for volunteers to come point to correct boxes as you present these sample scenarios:

✎ 7 children on a 5-cookie day. How many cookies in all? (The one box identifies the number of children, and the five box indicates the total number of cookies needed on a 5-cookie day.)

⑦	14	21	28	**35**
42	49	56	63	70
77	84	91	98	105

✎ 4 children on a 9-cookie day. How many cookies in all? (Again, the one box identifies the number of children, and the nine box indicates the total number of cookies needed on a 9-cookie day.)

④	8	12	16	20
24	28	32	**36**	40
44	48	52	56	60

- 9 children on a 3-cookie day?
- 10 children on an 8-cookie day?
- 7 children on a 12-cookie day?
- 12 children on a 6-cookie day?
- 5 children on a 13-cookie day?
- 3 children on a 14-cookie day?
- 11 children on a 7-cookie day?

Continue to call out combinations until students are able to locate the correct boxes without any difficulty or hesitation.

Internalize

Discover

This portion of the lesson involves a deliberate transition to the use of symbols. Use a set of multiplication cards with the answers printed on the back for this exercise. To begin, show the children the following problem on a flip chart or white board:

$3 \times 5 = 15$

1. Under the 3, draw 3 stick figures. Under the 5, draw five circles, or "cookies." To the right of the 15, draw 3 groups of 5 "cookies," like this:

2. Point to the 3 as you say, "3 children"; point to the 5 as you say, "times 5 cookies"; point to the 15 as you say, "is 15 cookies in all."
3. Say, "The short, easy way to say what is happening is 3 times 5 equals 15." (Point to each number in the equation as you say this.)

Practice

Materials needed:

Small counting items, such as plastic chips or small beans

Multiplication and division flash cards (one set for every pair or small group of students)

Have the children retrieve their master maps. As you conduct the activity, walk around the room and informally assess how each student is doing in his or her ability to locate an "address."

1. Call out a problem from your flash cards (for example, "7 × 8"). Check four students to see if they found the correct "address" and placed a counting item on that box.
2. Call out another problem. Check the work of the next four students.
3. Continue calling out problems and checking master maps to see if students are locating the correct answers.

Do this activity until your students are proficient at locating the correct answers to problems.

Challenge

Gather the students around you and your master map once more and pose this problem:

One day Mrs. Twig found 4 children around her table. When she checked her cookie supply, she discovered that she had 48 cookies. How could Mrs. Twig figure out how many cookies each child should get?

Let the children brainstorm for a bit and see what they come up with.

4	8	12	16	20
24	28	32	36	40
44	48	52	56	60

twelve box

✏ Did they identify the fours chart as the one to use?

✏ Did they locate the number 48 on that chart?

✏ Did they identify the twelve box as the one that indicated how many cookies each child would get? (The number 48 is found in the twelve box.)

✏ Restate the problem: "We have 4 children and 48 cookies. We want to divide the 48 cookies equally among the 4 children." (They each receive 12 cookies.)

✏ Do several more division problems together, making sure to emphasize that the chart used is determined by the number of groups involved.

Next, show them a card such as:

$$4\overline{)48}$$

Say, "There are 48 cookies inside the house, and 4 children are standing at the door, wanting to divide the cookies equally among themselves. We will need to divide 48 by 4."

Next, write this problem:

$$48 \div 4$$

Say, "48 divided by 4 is 12." Make the slicing motion with your hand as you say "divided by."

Master

✏ Practice with multiplication and division cards mixed up, then divide students into pairs or small groups and have them take turns reading problems. As one reads, the other locates the answer on his or her master map.

✏ Ask the children to explore the different locations for related problems such as $3 \times 5 = 15$; $5 \times 3 = 15$; $15 \div 3 = 5$; and $15 \div 5 = 3$.

Assess

Photocopy the chapter 4 assessment (T4.1, page 177). Tell students to read each problem and draw a line to the box that corresponds to each problem. Check for mastery, and reteach those who answered incorrectly in a small group, while the other students repeat the activity in pairs. As each child masters this subject, record evidence of that student's accomplishment on the Concept Mastery Tracking Chart.

Magnetic Patterns

Goals for This Chapter

1. Establish a visual background to assist in learning multiplication and division facts

2. Discover patterns that offer hooks for memory and recall

3. Gain a sense of the many patterns found in numbers

3. See numbers in relationship to each other, not as isolated entities

There is an order, a predictability, and almost a rhythm to the patterns that numbers make when arrayed in various ways. Patterns, once discovered, provide rules for the way in which they are arranged and make learning facts far simpler. The human brain is fascinated by patterns. When students are exposed to patterns, they are captivated. This chapter explores the patterns to be found in the master map.

Each student will need a few copies of the master map and colored pencils or crayons (or one laminated copy on which they can write with dry-erase markers). Coloring patterns helps to create a visual that students can recall as they internalize facts.

Introduce

Gather students around you and your large, laminated copy of the master map and tell them a story:

> During those evenings Mrs. Twig spent exploring her master map, she began to notice patterns in the numbers on her map. She hurried to get some colored pencils so she could color in some of the patterns she found. Before too long, Mrs. Twig's master map looked simply beautiful—and, best of all, she found that the colors on her map helped her remember where the numbers were found on it.
>
> Mrs. Twig noticed that one side of the master map featured only even numbers. Next, she noticed that the fifth column of the even side contained numbers that ended in zero. Mrs. Twig drew a line with her colored pencil to connect all the zeros.
>
> She noticed many other patterns as well. Let's see if we can find some patterns of our own.

Investigate

Have the children work with partners to see what patterns they discover. There will be patterns that appear on the master map, patterns that appear in each individual chart, and patterns that appear in the columns of tens places and ones places. Students may color the patterns and prepare to share their discoveries with the class. Remember that it is not effective to attempt to teach a pattern, as the greatest meaning is derived when students find patterns themselves and share their discoveries with others. Some patterns students may discover include the following:

The Ones-Place Columns

✎ Even boxes end in the same digit as you move down the column.

2	4	6	8	10
12	14	16	18	20
22	24	26	28	30

✎ On the odd side of the chart, the numbers in the first and third rows end in the same last digit.

1	2	3	4	5
6	7	8	9	10
11	12	13	14	15

✎ The ones-place digits in the twos chart and the twelves chart are identical (2, 4, 6, 8, 0).

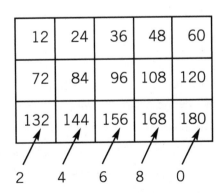

2	4	6	8	10
12	14	16	18	20
22	24	26	28	30
2	4	6	8	0

12	24	36	48	60
72	84	96	108	120
132	144	156	168	180
2	4	6	8	0

✐ The ones-place digits in columns 1 through 4 on the eights chart (8, 6, 4, 2) are in the exact opposite order as those in columns 1 through 4 on the twos and twelves charts (2, 4, 6, 8). Only the zeros remain in the same place on all three charts, "because the zero always stays home."

8	16	24	32	40
48	56	64	72	80
88	96	104	112	120

8   6   4   2   0

✐ The ones-place digits in columns 1 through 4 of the fours chart are 4, 8, 2, and 6. These numbers suggest a "back and forth hopping action."

4	8	12	16	20
24	28	32	36	40
44	48	52	56	60

4   8   2   6   0

6	12	18	24	30
36	42	48	54	60
66	72	78	84	90

6   2   8   4   0

✐ The ones-place digits in columns 1 through 4 of the sixes chart (6, 2, 8, 4) are in the exact opposite order as those in columns 1 through 4 on the fours chart (4, 8, 2, 6).

✐ The numbers in the tens chart all end in zero.

✐ The numbers in the fives chart all end in either 5 or 0, alternating like a checkerboard pattern.

The Tens-Place Columns

✐ On the evens side of the chart, each tens-place number increases by ½ of the chart number. For example, in the twos chart, as you move down the rows of a column, the tens-place number increases by 1 (½ of 2). In the fours chart, as you move down the rows of a column, the tens-place number increases by 2 (½ of 4).

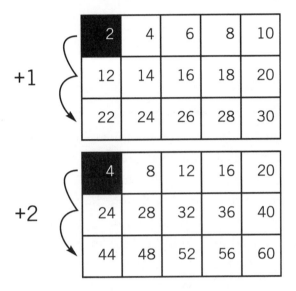

✐ If you compare the tens-place numbers in rows 1 and 3, you will find that the number in row 3 is the chart number written twice.

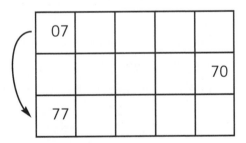

1. From 7 to 77 from box 1 to box 11
2. From 0 to 7 in the tens place

✐ Again, comparing rows 1 and 3 of each chart, you will find that the tens-place number in row 3 is exactly the amount of the chart number bigger than the tens-place number in row 1. For example, in the sevens chart, the tens-place number in the one box is 0, while the tens-place number in the eleven box is 7.

Whole Chart Patterns

✐ Note the pattern found in the "anchor numbers" on every chart. The anchor numbers make for a quick reference in determining the location of the other numbers.

Anchor Numbers

9				
				90
99				

✐ Anchor numbers serve as reference points for rapid recall of the numbers in each chart. Anchor numbers are found in the one, ten, and eleven boxes. The pattern rule for anchor numbers is as follows: the first anchor number, which is found in the one box, is itself (the chart number); the second anchor number, found in the ten box, is the chart number with a zero added to it; and the third anchor number, found in the eleven box, is the chart number written twice.

✐ Choose any box number in the ones chart and read the numbers in that same box in each subsequent chart. You will find yourself chanting that box number's times table. For example, if you go to the two box in the ones chart, then go to the two box in each of the following charts, you will be counting by twos.

✐ Compare the number found in the five box of each chart on the evens side of the master map to the number found in that chart's one box (the chart number). Looking at the twos chart, see that the number 10, which is located in that chart's five box, is exactly ½ the chart number, but with a zero added to the right of the number.

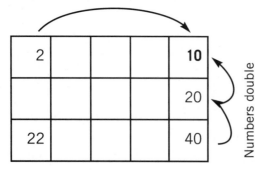

Columns:

<table>
<tr><th colspan="5">Odds</th><th colspan="5">Evens</th></tr>
<tr><th>1</th><th>2</th><th>3</th><th>4</th><th>5</th><th>1</th><th>2</th><th>3</th><th>4</th><th>5</th></tr>
</table>

Rows:

1	2	3	4	5	2	4	6	8	10
1 1	2	3	4	5	2	4	6	8	10
2 6	7	8	9	10	12	14	16	18	20
3 11	12	13	14	15	22	24	26	28	30

3	6	9	12	15	4	8	12	16	20
18	21	24	27	30	24	28	32	36	40
33	36	39	42	45	44	48	52	56	60

5	10	15	20	25	6	12	18	24	30
30	35	40	45	50	36	42	48	54	60
55	60	65	70	75	66	72	78	84	90

7	14	21	28	35	8	16	24	32	40
42	49	56	63	70	48	56	64	72	80
77	84	91	98	105	88	96	104	112	120

9	18	27	36	45	10	20	30	40	50
54	63	72	81	90	60	70	80	90	100
99	108	117	126	135	110	120	130	140	150

11	22	33	44	55	12	24	36	48	60
66	77	88	99	110	72	84	96	108	120
121	132	143	154	165	132	144	156	168	180

39

✏ On the odds side of the master map, the number found in the five box is equal to the smaller half of the numbers that add up to the chart number, with a five beside it. For example, in the threes chart, the chart number is three. One plus two equals three. The smaller "half" of one and two is one. Now put a five beside the one, and you get fifteen, or the number in the five box.

3 = 1 + 2						5 = 2 + 3				
3				**15**		5				**25**
				30						50
33				45		55				75

Internalize

Discover

Gather the class together as a group and discuss what you have found. These explorations of patterns strengthen students' "noticers" and help children develop a habit of finding their own learning clues—those wonderful details that provide hooks for memory and recall. Allow time for, and encourage, students who are not sharing at any particular moment to make note of patterns their classmates found that they did not.

Practice

Have students go back to their partners with a copy of blackline master S3.1 (page 98).

✏ Have them practice filling in the anchor numbers of each chart.
✏ Have them practice locating and writing down the patterns found in the ones-place column of each chart (for example, the ones places in the twos and twelves charts are alike, the ones places in the eights chart are just backward to those, the ones places in the fours and sixes charts are backward from each other, and so forth).

✐ Practice writing in the tens-place columns on the evens side of the master map.

✐ Practice filling in, at random, the number that belongs in the five box of each chart.

Students will likely rise to the challenge of finding more "rules," or patterns, within the master map. Those students who want more challenge may search out additional patterns. The more all students study the master map, the more they will gain in terms of identifying the locations of the answers to their multiplication and division problems.

Master

Give students sufficient time to study the master map in their pairs, and encourage them to check each other's maps for accuracy. Give them the freedom to choose which box numbers they will check. For example, one day a pair might decide to practice anchor numbers. For each practice session, students will need a copy of blackline master S3.1 (page 98).

Assess

The goal of any assessment at this stage is to determine which students are engaged in the process of pattern discovery and which students are not. You can gain valuable information about each child as you make an informal assessment of his or her work. Take notes regarding each child's performance, checking for the following:

✐ What is the student's level of engagement? Is the student excited, somewhat interested, or neutral?

✐ Is the student actively participating, or is he or she letting the partner do the "discovering" for both of them? When pairing up students, avoid pairing a retiring child with a strongly verbal one. The activities will be of value as each child is contributing ideas.

✐ How significant are the student's findings in terms of providing memory hooks? For some children, these pattern rules will make a difference in their ability to remember numbers, while for others, the gains might be marginal. Make a note of these differences, as it will help you understand each student's learning style.

✏ How is the student doing when it comes to finding new patterns? Those who forge ahead, determined to find something new, might be the children who are the strongest risk takers in terms of seeking connections beyond what the lessons provide—the students who are good at making personal connections for learning.

6

Make It Happen

Goals for This Chapter

1. Create an environment that promotes learning

2. Explore a sample lesson for learning tables on the master map

Now that the foundation for learning multiplication and division is in place, it is time to finish the job. Two important factors will make the remainder of the task successful: the creation of a proper environment for learning math and the procedure used to teach each individual chart. The steps of the procedure are the same for each chart; once you are familiar with the sequence of this sample lesson, the task of teaching each chart will be easier.

Create an Environment for Learning Math

Following are some important considerations for creating an environment that will maximize learning potential:

Physical Environment

- **Materials.** Teachers often underestimate the importance of the visual appearance of materials that are presented to students. Photocopied handouts are sometimes crooked on the page or faded. The benefit of a visually appealing page is worth the added effort and time required to make good, easy-to-read photocopies. For work to be done in color, connive to secure new boxes of crayons, colored pencils, or markers.

- **Music.** Consider playing background music during students' study of math and later during your assessment of their progress. Much has been written about the effects of some types of music on the thought patterns of children who are studying. At a minimum, the music will soothe students; at best, it will enhance learning and recall.

- **Lighting.** Soft lamplight helps to provide a welcoming environment for students.

- **Learning styles.** Pay attention to students' individual learning styles. Some children are much better able to focus when sprawled on a piece of carpet and given a clipboard to write on, while others focus better when they are seated at their desks.

- **Student placement.** Take notice of which students do best while studying alone, and which profit greatly from a study partner. Students should be taught that the goal of accommodating individual learning preferences is to ensure that each child can focus to the best of his or her ability. If a student chooses to work with a partner, but ends up chatting with that person instead of working, it's best to place one or both of them in a different setting that enhances their ability to focus.

- **Cold water.** Have cold water available for students to sip while working. Cold water can help kids wake up, stay alert, and maintain their focus. It is also commonly known that the brain requires large amounts of oxygen and water for optimal functioning.

- **Use color.** Encourage highly visual children to color code number patterns. Seeing colored patterns on a master map might prove to be another way for a visual learner to remember the arrangements of numbers. In order for this approach to be of help, however, the students themselves must be interested in using color, and each must be allowed to color code their own maps as they desire.

Emotional Environment

- **Expectancy.** Model an air of expectancy when you approach the study of math. There is majesty in the number arrays, and if your own fancy is caught by the subject, you will convey by your very demeanor that "this is the part of the day that we've been waiting for." Students will mirror your attitude.

- **Movement.** Consider doing jumping jacks prior to starting the lesson. Not only does this increase the oxygen supply available to the students' brains, but also, if you tell them that they are doing warm ups for the task ahead of them, students are inclined to view the task of learning math as important.

- **Collaboration, not competition.** Share with the class the goal of every student reaching 100 percent mastery of the master map. Enlist their help in making this happen. Model the qualities of helpfulness and collaboration, and praise students who take the initiative to offer help to a fellow classmate.

- **Metacognition.** Your priority as a teacher should be on encouraging those strategies that increase focus for students. By emphasizing this goal, students will follow your example, and they will learn to pay attention to those elements that increase their own ability to focus and promote their own learning.

Following are some ideas to consider with children who have difficulty in focusing:

✐ **Solo work.** Students who have attention challenges may require a seat with no view of other children. Ask these students if working solo might help them to focus.

✐ **Task list.** Some students benefit from having a short written list of tasks to accomplish. As tasks are completed, students can check them off their list.

✐ **Allies in learning.** The goal with easily distracted children is to make them allies in the quest for focus. Ask them to pay attention to what helps them stay focused. For some students, using their hands in some way (such as writing notes, coloring patterns, or arranging numbers within a grid) is imperative when working on a task that requires focus.

✐ **Self-monitoring.** Some children benefit from being instructed to not look up from their papers until they have completed a defined, small amount of work, such as two rows of a chart. After looking up, they must do two more rows.

✐ **Use the pointer finger.** Have easily distracted students use their pointer fingers to direct their brains to the object of their focus. For example, have them start a written test or activity by pointing to problem 1, and by keeping their pointer finger there while their other hand writes the answer. Have them then immediately point to problem 2 so that their minds are immediately directed to that problem. Instruct students to continue in this fashion until they have completed all of the problems on the page. If necessary, a simple, "Where is your pointer?" reminds these students to stay focused on the next problem.

Sample Lesson—Teaching Multiplication and Division Charts

1. **Seek patterns.** Review patterns that you discovered in previous lessons. Review anchor numbers, magic center numbers, and number sequences discussed in chapter 5.

2. **Use color.** Color each chart (if desired by students). Reinforce pattern discovery by noting the placement and arrangement of the colors on the chart.

3. **Identify clues.** Discover a defining clue for each chart. This clue must be obvious enough to provide immediate recall of the numbers placement on the chart.

4. **Individual practice.** Give students copies of the empty grids found on page 100 (blackline master S6.1) and have them practice filling them in. Be creative in your instructions on filling out the charts. For example, you might have students fill in anchor numbers, then numbers that you randomly select. You might have them fill in the ones-place columns first, then fill in the tens-place columns. The goal is to avoid having them simply start with the one box and count up sequentially. Repeat this practice three to six times. Some children will achieve a level of fluency by the third time they fill in the chart, while others will need further practice before the numbers click with them.

5. **Partner practice.** Have students work in pairs and take turns being the "questioner" and the "answerer." As the questioner points to a specific box on the empty grid, the answerer should write the appropriate product/quotient in that box. During the next round, have the questioner call out a box number, and have the answerer write the correct number that is associated with that box. In the third round, have the questioner call out a number, and have the answerer write that product/quotient in the correct box.

6. **Multiplication practice.** Give the children copies of the multiplication worksheet to which each lesson directs you. Each worksheet contains several problems and a small, empty grid. Instruct students not to write the answers in the grid, but to use the empty grid as a means for locating a product from inside the chart. Their brains will actually supply an answer when their index fingers point to the box that "holds" the answer to a problem they are solving. For example, when solving the problem 9 x 7, a student would point to the seven box on the empty chart and "see" the number 63 within the pattern on the grid. Students are able to do this because they have become familiar with where each product fits into the patterns particular to each chart.

7. **Tie multiplication to division.** Use the Multiplication and Division Connection sheets found in this book to reinforce the relationship that exists between them. Next, repeat step 6, but this time with division worksheets, using the masters as directed in the individual lessons. Again, do not allow the children to write the answers on the empty grid. If they need to, they may look at the grid to determine which box the dividend came from, which should allow them to "see" the quotient in their mind's eye.

8. **Mixed practice.** Next, provide copies of the worksheet called for in the individual lessons and have the children practice with a combination of multiplication and division problems for the chart that is the subject of the lesson.

9. **Assign homework.** Have students complete a practice sheet as directed in the lesson. This sheet is identical to the sheet used in step 8. The purpose of repeating this exercise is to allow students extra practice.

10. **Review.** On the following day, combine the chart you are studying with problems from a previous chart. (If you are working with the first chart, go on to a new lesson and repeat the steps 1–9.) Mixed practice worksheets are provided for each lesson.

Tips to Remember

- Review frequently but briefly. Success will result from a few minutes' practice each day.
- Allow students to take charge of their learning process. Rely on each student to tell you when he or she is in need of additional practice.

Assess

Assessments consist of the pages listed in each lesson in chapter 7. Do not time assessments.

- As students complete these assessments, informally evaluate their progress. Notice the way in which students complete their problems.
- Be vigilant in spotting students who count up to reach a target number. Discourage this practice, as it is quite limiting. Once children learn to rely on counting up to reach an answer, they will always select that same strategy in order to arrive at an answer. Although counting up usually results in a correct answer, it's a slow way of going about the task. In addition, children gain no understanding of patterns and relationships that exist between numbers when they simply count sequentially.
- Record each child's mastery of each chart on the Concept Mastery Tracking Chart (T2.3, page 175).

Make Memories

Goals for This Chapter

1. Discover patterns for each chart

2. Provide a connection between multiplication and division

3. Practice completing each chart

4. Prove new knowledge by completion of problem sets

5. Practice target problems for multiplication and division

6. Combine new learning with prior learning

Begin the memorization of charts on the master map by targeting the even-numbered charts first. This is because of the similarities between the patterns on the evens side of the master map. In this chapter, each chart is addressed and should be presented to students as per the lesson sequence presented in chapter 6. The order in which the even-numbered charts are addressed is as follows: 10, 2, 12, 8, 4, and 6. The odd-numbered charts sequence is 11, 5, 9, 3, and 7. The reasoning behind these orders is explained in this chapter.

Even-Numbered Charts

Tens Chart

1. **Seek patterns.** As an anticipatory activity, discuss number patterns that the children discover in this chart. Two obvious patterns are that every ones-place number is a zero, and that the tens-place numbers rise sequentially and match the box numbers exactly. Each number, therefore, in the tens chart is the box number followed by a zero.
2. **Use color.** Color the chart if students find this activity to be helpful.
3. **Clues.** One clue that identifies the tens chart is the sequence of box numbers with the added zero following, or "box number and zero."
4. Follow steps 4 through 9 of the sample lesson (page 46) using copies of blackline masters S7.1, S7.2, and S7.3 (pages 101–103). (You will not have a step ten for this first lesson because it is the first chart we are studying.)

Twos Chart

1. **Seek patterns.** Discuss number patterns that the children discover in this chart. In the twos chart, the ones-place numbers follow this sequence down the columns:

2	4	6	8	10
12	14	16	18	20
22	24	26	28	30

Counting by twos in ones-place columns.

In the tens places, there are four (implied) zeros, then five ones, then five twos, then one three.

0, 0, 0, 0, 1
1, 1, 1, 1, 2
2, 2, 2, 2, 3

2. **Use color.** Color the chart if students find this activity to be helpful.

3. **Identify a chart clue.** The defining clue for the twos chart is the sequence two, four, six, and eight in the ones place. We could say, "Count by twos" for short.

4. Follow steps 4 through 9 of the sample lesson (page 46) using copies of blackline masters S7.4, S7.5, and S7.6 (pages 104–106). A day later, review the defining clues for both the twos and the tens chart, then assess students' mastery of these charts using copies of blackline master S7.7 (pages 107–108). Pay attention to any problems that crop up at this point, and use the problems as opportunities to review or reteach.

Twelves Chart

1. **Seek patterns.** The twelves chart is next because it has the same array of twos multiples in the ones place as the twos chart. In the tens places, numbers increase sequentially, except that there is a one-digit gap between columns four and five. Thus the numbers read like this:

5, 11, 17 missing here

12	24	36	48	60
72	84	96	108	120
132	144	156	168	180

2. **Use color.** Color the chart if students find this activity to be helpful.

3. **Identify a chart clue.** The defining clue of the twelves chart is counting by twos in the ones places. Another clue is the three disappearing tens-place

numbers between columns four and five. We could say, "Count by twos, three lost numbers."

4. Follow steps 4 through 9 of the sample lesson (page 46) using copies of blackline masters S7.8, S7.9, and S7.10 (pages 109–111). Record each student's mastery of the twelves chart on the Concept Mastery Tracking Chart.

Assess—Halfway Point

At this point, having gone through three of the six even-numbered charts, review the tens, twos, and twelves charts. Then give students copies of blackline master S7.11 (pages 112–114), which contains all the problems found on the worksheets for the three charts. Check for mastery. Pay attention to which problems, if any, the students answer incorrectly. Are all (or most) students getting the same problem or problems wrong? If so, review or reteach the concept involved in that problem. Provide daily reviews.

Eights Chart

1. **Seek patterns.** Pattern discovery will reveal that the color bars (the "count by twos" columns) are still with us, but this time they appear in exactly the opposite order as in the twos and twelves charts. In the eights chart, the twos bars march backward like this: 8, 6, 4, and 2. Notice also the tens-place numbers found in the five box and in the six box, as well as in the ten box and the eleven box.

8	16	24	32	40
48	56	64	72	80
88	96	104	112	120

Repeating tens in 5 & 6, 10 & 11 boxes.

Only the tens-place numbers four and eight repeat in this chart. All the other numbers that are found in the tens place appear only once.

2. **Use color.** Color the chart if students find this activity to be helpful.

3. **Identify clues.** Two defining clues for this chart are the twos multiples in the ones places decreasing sequentially from one column to the next and the

sequential tens-place numbers, double fours and double eights. Or we could say, "Countdown, double 4 and 8."

4. Follow steps 4 through 10 of the sample lesson (page 46) using copies of blackline masters S7.12, S7.13, and S7.14 (pages 115–117). Use copies of blackline master S7.15 (pages 118–121) for step 10, in order to provide more practice and to assess the areas that might need to be retaught.

Fours Chart

1. **Seek patterns.** Note that the color bars still appear, but now they have left their original sequence and have made a new one. Now the color bar sequence looks like this:

4   8   2   6

This sequence of numbers also forms a sentence: "4 doubled = 8, – 2 = 6." Kinesthetic learners will enjoy the bouncing motion of the new number arrangement. Starting at the number two, they can bounce to the left to hit the number four, bounce to the right to hit the number six, then bounce back to the left to hit the number eight.

	4	8	1 2	1 6	②0
+2 | ②4 | ②8 | 3 2 | 3 6 | ④0 |
+2 | ④4 | ④8 | 5 2 | 5 6 | 6 0 |

Tens sequence: 0, 0, 1, 1, 2, 2, 2, 3, 3, 4, 4, 4, 5, 5, 6.

In the illustration above, note that the tens-place sequence is consistent except when it comes to the twos and fours, which appear three times instead of twice. Review the anchor numbers. In the 5–6 box sequence, the tens-place number in the five box is ½ of the chart number followed by zero, while in the six box, the tens-place number is ½ of the chart number followed by the chart number itself. Compare this discovery to the numbers found in the ten and eleven boxes.

2. **Use color.** Color the chart if students find this activity to be helpful.

3. **Identify clues.** The 2/4 triplets pattern (20, 24, 28; 40, 44, 48) provides one hook for remembering this chart, as does the saying (for the ones-place numbers in columns) "4 doubled = 8, – 2 = 6."

4. Follow steps 4 through 10 of the sample lesson (page 46) using copies of blackline masters S7.16, S7.17, S7.18, and S7.19 (pages 122–128).

Sixes Chart

1. **Seek patterns.** Note that the arrangement of numbers in the ones places in the first four columns is the exact opposite of the arrangement of numbers in the ones places in the first four columns of the fours chart. Here, the sequence is six, two, eight, four. Note also that there are two of every tens-place number except for the numbers that appear in column four (two, five, and eight) and in boxes 1 and 15. Review the anchor numbers, the 5–6 sequence, and so forth.

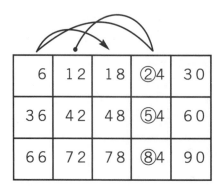

6	12	18	②4	30
36	42	48	⑤4	60
66	72	78	⑧4	90

Circled numbers appear only once.

2. **Use color.** Color the chart if students find this activity to be helpful.

3. **Identify clues.** The defining clue of the sixes chart is "6 + 2 = 8, halved = 4."

4. Follow steps 4 through 9 of the sample lesson (page 46) using blackline masters S7.20, S7.21, S7.22 (pages 129–131), and S7.23 (pages 132–135). Then conduct a general review of the evens charts of the master map.

Even-Numbered Charts Assessment and Tips

Use copies of blackline masters T7.1, T7.2, and T7.3 (pages 178–183) to assess students' mastery of all six even-numbered charts.

✐ Remember to conduct informal kid-watching assessments rather than relying solely on formal, written assessments.

✐ Frequently ask students "How did you remember that one?" and "How can you remember that one?" The first question calls children's attention to their own thinking processes, and the second question enlists the brains of students in making their own connections.

✐ Remember that frequent practice is essential to the acquisition of long-term memory. Make extra copies of assessment sheets and have them available for students to use when you give them five minutes or so for practice. Let students choose which practice sheet to use.

Odd-Numbered Charts

Elevens Chart

1. **Seek patterns.** The discovery of patterns in the elevens chart is a pure delight. Double numbers march across three rows, with only one small variation in row 3, where the number found in each box is split in the middle by a third digit that is one digit larger than the ones-place number found in that box. The center digit is also the sum of the two outer digits.

Third row number patterns.

2. **Use color.** Color the chart if students find this activity to be helpful.
3. **Identify clues.** The defining clues for the elevens chart are the doubling-up, or "stuttering," of the box-number digit, and in row three, the larger center digit. The elevens chart may be referred to as "the stutter, jump back chart."
4. Follow steps 4 through 10 of the sample lesson (page 46) using blackline masters S7.24, S7.25, and S7.26 (pages 136–138).

Fives Chart

1. **Seek patterns.** The fives chart features a very orderly array of numbers. The ones-place numbers are all either zeros or fives, and these digits alternate

in a checkerboard pattern. As for the tens-place numbers, there are two of every digit except for the (implied) zero in box 1. The tens-place number pairs are all found sitting side by side except for the tens-place number five, which is in boxes 10 and 11. Review the anchor numbers, the magic center, the 5–6 box sequence, and the rule for the fifth column, which is to add the number in the five box and the number in the ten box, which equals the number in the fifteen box. (This rule applies to every chart on the master map.)

5	1⓪	1 5	2⓪	2 5
3⓪	3 5	4⓪	4 5	5⓪
5 5	6⓪	6 5	7⓪	7 5

Checkerboard ones places.

2. **Use color.** Color the chart if students find this activity to be helpful.
3. **Identify clues.** One defining clue of the fives chart is its checkerboard pattern of fives and zeros in the ones places. Another defining clue is its pattern of pairs of tens numbers, which is arrayed in a zigzag fashion. The fives chart can be referred to by the phrase "checkerboard and zigzag."
4. Follow steps 4 through 10 of the sample lesson (page 46) using copies of blackline masters S7.27, S7.28, S7.29, and S7.30 (pages 139–143).

Nines Chart

1. **Seek patterns.** The nines chart contains many patterns. For example, the sum of the digits in each box is nine. In addition, the ones-place numbers decrease sequentially from nine to zero in the first two rows, and the countdown starts all over again with the third row.

9	8	7	6	5
4	3	2	1	0
9	8	7	6	5

When the three-digit numbers appear, the first two digits increase sequentially by one (10, 11, 12, 13), and the ones-place numbers decrease by one from right to left:

Counting up . . .

108   117   126   135

Counting down . . .

2. **Use color.** Color the chart if students find this activity to be helpful.

9	18	27	36	45
54	63	72	81	90
99	108	117	126	135

3. **Identify clues.** The defining clue for the nines chart is surprising: for any number times nine, take that number (the box number) reduce it by one, and add a digit that would make the box number total nine.

$$-1 \quad \begin{array}{r} 9 \\ \times 3 \\ \hline 27 \end{array} \quad -1 \quad \begin{array}{r} 9 \\ \times 4 \\ \hline 36 \end{array} \quad -1 \quad \begin{array}{r} 9 \\ \times 5 \\ \hline 45 \end{array} \quad -1 \quad \begin{array}{r} 9 \\ \times 6 \\ \hline 54 \end{array}$$

$$2+7=9 \qquad 3+6=9 \qquad 4+5=9 \qquad 5+4=9$$

Other clues are that the eleven box contains a simple stutter of digits: 99; and in boxes 12–15, the multiplier is reduced by two instead of one, and then a digit is added to equal nine as before.

$$-2 \quad \begin{array}{r} 9 \\ \times 12 \\ \hline 108 \end{array} \quad -2 \quad \begin{array}{r} 9 \\ \times 13 \\ \hline 117 \end{array} \quad -2 \quad \begin{array}{r} 9 \\ \times 14 \\ \hline 126 \end{array} \quad -2 \quad \begin{array}{r} 9 \\ \times 15 \\ \hline 135 \end{array}$$

$$1+0+8=9 \quad 1+1+7=9 \quad 1+2+6=9 \quad 1+3+5=9$$

Reduce multiplier by 2, then add a digit to equal 9.

When dividing, the defining clue is even simpler. The answer to any nines division problem is the digit that is one larger than the tens-place digit of the dividend. Look at the problems below:

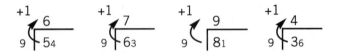

With triple-digit dividends, the answer increases by two digits instead of one:

4. Follow steps 4 through 10 of the sample lesson (page 46) using copies of blackline masters S7.31, S7.32, S7.33, and S7.34 (pages 144–148).

Threes Chart

1. **Seek patterns.** The threes chart features the partner numbers together again, but the numbers have moved from their original positions to explore new territory. Each of the tens-place numbers appears three times except for the number 3, which appears four times, taking the place of one of the fours (box 13). It is, after all, the threes chart!

3	6	9	12	15
18	21	24	27	30
33	36	39	42	45

Notice the three numbers that contain the number 2 in the tens place (21, 24, and 27). These numbers form the center of the threes chart.

Review the 5–6 sequence, the anchor numbers, and the fifth column sequence. In this chart, just as in the sixes chart, two pairs of numbers mirror: 24 and 42, 12 and 21.

2. **Use color.** Color the chart if students find this activity to be helpful.

3. **Identify clues.** The defining clues for the threes chart are: the tens-place numbers are tripled (000, 111, 222, 3333, 44—extra 3); the center twenties numbers (21, 24, and 27); and the zigzag twos. The threes chart may be referred to as, "Triplets, twenties, and zigzag."

4. Follow steps 4 through 10 of the sample lesson (page 46) using copies of blackline masters S7.35, S7.36, S7.37, and S7.38 (pages 149–153).

Sevens Chart

1. **Seek patterns.** In the sevens chart the tens-place numbers increase sequentially, but there are double twos, fours, and nines. Note also that the ones-place partner numbers (1 and 6, 2 and 7, 3 and 8, 4 and 9) are each clustered together in the same columns. This pattern occurs in every odd-numbered chart.

7	14	21	28	35
42	49	56	63	70
77	84	91	98	105

7	14	②1	②8	3 5
④2	④9	56	63	70
77	84	⑨1	⑨8	105

2. **Use color.** Color the chart if students find this activity to be helpful.

3. **Identify clues.** The defining clues for the sevens chart are the "zigzag twos" design and the positioning of the tens-place number pairs (2, 2; 4, 4; 7, 7; 9, 9).

4. Follow steps 4 through 10 of the sample lesson (page 46) using copies of blackline masters S7.39, S7.40, S7.41, and S7.42 (pages 154–158).

Assessment and Tips

✐ Use copies of blackline masters T7.4 and T7.5 (pages 184–187) to assess students' mastery of odd-numbered charts.

✐ For formal assessments, use assessments T7.6, T7.7, T7.8, and T7.9 (pages 188–193). Use the assessments to gain a sense of student mastery, as well as to pinpoint areas that need particular review. Record each student's mastery on the Concept Mastery Tracking Chart.

✐ Remember that fluency is gained with the building and growing of dendrites—those tiny connectors in the brain. The more children practice, the more fluid and automatic their work becomes.

✐ Work for mastery.

✐ Keep extra copies of assessment sheets available to students so that they can review what they've learned by completing a sheet from time to time. Try having students review twice a week at first, then once a week as fluency is gained. Of course, once the work is automatic, further review will not be needed.

✐ Practice the habit of conducting a quick (five minutes or so) review of a particular chart or pairs of charts at the beginning of each math class.

Multi-Digit Multiplication

Goals for This Chapter

1. Activate prior knowledge of place values and times tables

2. Experience the action of multiplication

3. Create a framework for multiplication by using a story

4. Relate the story of multiplication to place values

5. Practice multi-digit multiplication to achieve mastery

Stories are powerful tools for teaching new or difficult concepts. Beginning a lesson with a story takes the threat out of the difficult concept and eases your now-receptive students into learning. By using real materials to model the action of a new concept, you show students what is really happening when they solve the problems using symbols. Their understanding of the action will provide a link between concept and memory.

Introduce

Gather students around you. For this lesson, you will need a flip chart or white board, a set of markers, and a model of three connected buildings (see T8.1, page 194). Tell students the following story:

> Let's go back and visit Mrs. Twig and find out what has been happening to her. Do you think she's still making cookies? Or do you think she has moved on to something new? (Let the children predict what Mrs. Twig might be up to these days.)
>
> Mrs. Twig was still making cookies! As a matter of fact, her cookies were so good that more and more people started to visit her kitchen. So many people came, in fact, that her kitchen was bursting at the seams. Mrs. Twig thought hard about what to do. Finally, she decided to walk into town to see if she could find a building that was bigger. She wanted a place with enough room for big, new ovens, and she needed a lot of room for the people who came to visit her.
>
> After walking around for a while, Mrs. Twig happened upon a perfect building. It had three rooms that were connected by doors, and in the rooms were shelves and counters and big windows. Mrs. Twig was delighted! She quickly moved in. Before long, people were flooding into her building to buy cookies. (Draw the children's attention to the drawing of the building.)
>
> In the first room, Mrs. Twig sold cookies to people who just wanted a few—say, 9 or fewer. These she would just bag up and hand to her customers. Many children came into this first room!
>
> In the middle room, Mrs. Twig sold boxes of cookies. Each box contained 10 cookies. (Point out the middle room, and particularly the number 10 that is written in the attic of that room).
>
> Sometimes people would come to Mrs. Twig's cookie shop to buy a great many cookies. Some of these people owned restaurants; some were

having big parties or weddings, and they needed really big boxes of cookies. In the third room (point this room out to your students), Mrs. Twig sold boxes that contained 100 cookies each.

One morning, a man hurried in the door to find Mrs. Twig. He seemed to be distressed. He had a problem! Mrs. Twig was always glad to talk to her customers and help them solve their problems. By this time, she had learned all the charts on her master map, so it was not difficult at all for her to figure out how many cookies her customers needed.

"How can I help you, Mr. Fiffle?" asked Mrs. Twig.

"Oh, Mrs. Twig! I am having a party for 45 people, and I want to give each of them 6 cookies. I don't know how many boxes of cookies I need to buy!"

"Don't worry, Mr. Fiffle," said Mrs. Twig. "I can help you with that!" She drew a picture that looked like this:

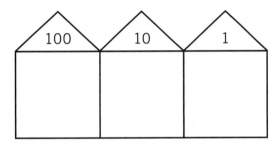

Mrs. Twig said, "You have 45 people coming. Let's see. That is 4 groups of 10 people, plus 5 more people." As she said this, Mrs. Twig wrote the number 4 in the tens room and the number 5 in the ones room, like this:

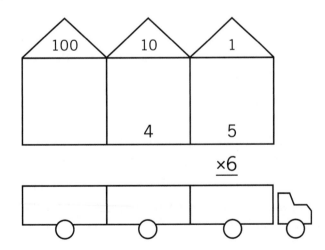

"If you were going to give each guest 1 cookie, you could just buy 4 boxes of 10 cookies and 5 single cookies," said Mrs. Twig, "but you are not doing that. You want each guest to have 6 cookies." As she said this, Mrs. Twig wrote a "x 6" under the drawing of her shop. She also drew in the delivery truck that would carry away Mr. Fiffle's cookies.

Procedure

1. Mrs. Twig decided to start with the 5 single people first. She knew that 6 × 5 = 30, so she knew immediately that she would not need any single cookies from the ones room. Mrs. Twig wrote a zero on a piece of cardboard and hung it on a hook on the wall in the ones room. Then she hurried into the tens room. She wrote the number 3 on a piece of cardboard and hung it on a hook in that room. The numbers on the pieces of cardboard would remind her how many boxes of cookies she needed to take from each room.

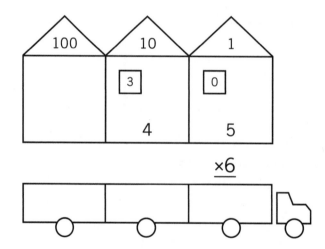

2. (Recap by saying, "For the 5 single people, we will need 3 boxes of 10 cookies and no single cookies.")
3. Next, Mrs. Twig turned her attention to the number 4 that was written in the tens room. "We have the cookies we need for the 5 single people," she said. "Now let's figure out how many cookies we need for these 4 groups of 10 people."
4. Mrs. Twig said, "I will multiply the 4 10s times 6 to get the answer: 24 tens. But that's a problem! I'm in the middle room, and I sell no more than 9 boxes of 10 cookies each in this room."

5. In order to solve this problem, Mrs. Twig wrote the number 4 on a card and hung it with the number 3 that she had already hung up on the wall. Then she hurried into the hundreds room, quickly wrote the number 2 on another card, and hung the card on the wall in that room.

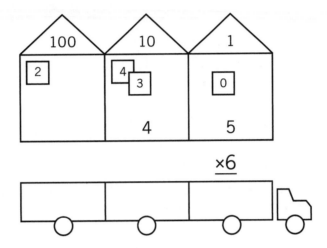

6. (Recap by saying, "In order to feed 45 people 6 cookies each, we will need 2 boxes of 100 cookies, 3 + 4, or 7, boxes of 10 cookies, and no single cookies. Let's load the truck!")

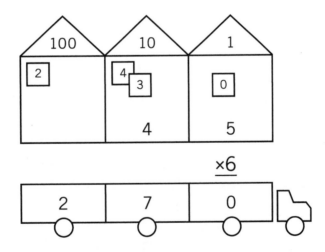

7. Mrs. Twig put no cookies in the ones part of the truck. She put 7 boxes of 10 cookies in the tens part of the truck. Finally, she put 2 boxes of 100 cookies in the hundreds part of the truck.

8. (Recap by saying, "We need 270 cookies for this party.")

9. (Walk the children through the problem again, leaving the original drawing up on the white board. Show this problem:)

$$\begin{array}{r} 45 \\ \times 6 \\ \hline \end{array}$$

10. (Say, "6 x 5 = 30. 3 tens and 0 ones." Hang the 3 tens on the wall in the tens side, and the 0 in the ones side.)

$$\begin{array}{r} 30 \\ 45 \\ \times 6 \\ \hline \end{array}$$

11. (Say, "6 × 4 tens = 24 tens. Write the number 2 in the hundreds room, and the number 4 in the tens room.)

$$\begin{array}{r} 24 \\ 30 \\ 45 \\ \times 6 \\ \hline \end{array}$$

12. (Say, "Load the truck!" As you load the truck ask, "How many ones?" [0], "How many tens?" [7], and "How many hundreds?" [2].)

Next, practice a problem that involves a different set of numbers—say, 69 × 9.

1. Using a clean picture of the bakery rooms, write the numbers 6 and 9 in the appropriate rooms.

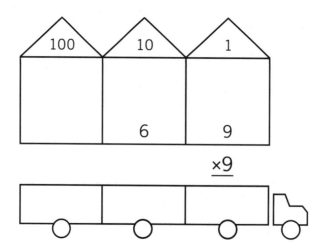

2. There are 69 people, and each will receive 9 cookies.

3. Multiply the singles first: 9 × 9 = 81, or 8 tens and 1 one.

4. Write these numbers on "cards" placed in the appropriate rooms in the picture.

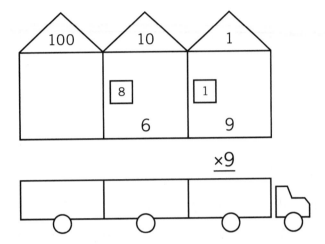

5. Next, multiply 6 tens x 9. The answer is 54 tens. The 4 tens may be sold from the middle room, but the 50 tens must be sold from the hundreds room.

6. Write the number 4 on a "card" and "hang it" in the tens room, then hurry next door to the hundreds room. Write the number 5 on a "card" and "hang it" in that room.

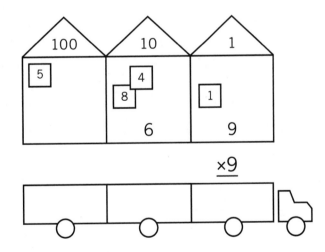

7. See if students notice that there are now too many tens hanging in the tens room. Load the truck!

8. How many ones? (1)

9. How many tens? (12. Oops! Too many!)
10. Load 2 boxes of 10 cookies into the truck, but carry a one next door to put with the five there.
11. Load 6 boxes of 100 cookies.
12. Ask, "Who noticed that Mrs. Twig had to make a 100 out of the 12 tens?" Remember that you cannot have more than nine of a kind in each room. 12 × 10 = 120, or 1 hundred and 2 tens.

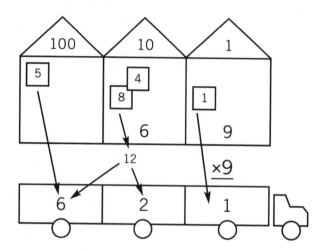

Load the truck.

Practice to Develop Fluency

Work through several more problems that involve the multiplication of one two-digit number and one one-digit number (such as 4 × 57; 3 × 68; 7 × 24; and 9 × 47). Have students supply answers orally while you write. When students are comfortable with the action in the story of multi-digit multiplication, give them each a copy of blackline master S8.1 (page 159) and have them solve the problems on that worksheet. You may have students work alone or with a partner. The goal at this point is not for students to arrive at the correct answers, necessarily, but to practice the action of multi-digit multiplication.

When they seem to understand the procedure, give students copies of blackline master S8.2 (page 160) to work on. This sheet provides a transition from pictures and symbols to symbols only.

Next, gather the class together again and work on this problem together: 5 × 49.

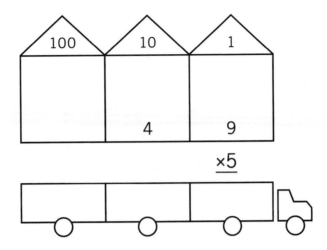

1. First, multiply 5 x 9 to get 45, or 4 tens and 5 ones.

2. Ask students, "Do you think we can go ahead a load the 5 ones now?" Let them discuss this idea to see what they determine. You want them to discover that the ones number will not change during further computation, so there is no reason why they cannot go ahead and load the 5, then write the number 4 on a card in the tens room.

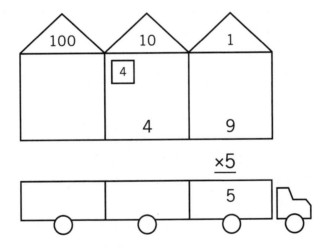

3. Next, multiply 5 by the 4 tens to arrive at 20 tens.
4. Note that 20 tens = 2 hundreds + 0 tens + 0 ones.
5. Write the 2 hundreds in the hundreds room.
6. Load the truck!

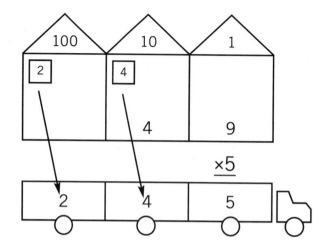

7. As an additional challenge, ask students if you could have gone ahead and loaded the boxes of 100 cookies as soon as you found that you were going to have two hundreds boxes. Let them discuss this issue among themselves. See if they conclude that the hundreds digit would not change, so it could be "loaded" as you were working out the problem. Don't push this issue. Students need to discover this shortcut themselves as they are solving problems. Eventually, they will notice that they can eliminate some steps. You do not want them to memorize steps, rules, or shortcuts, with no real understanding of what they're doing. The story will guide them to understand the steps to take, and practice will show them what shortcuts they can take.

Challenge

1. Ask students how they would solve a problem that involved a triple-digit number, such as 3 × 231. (In this particular problem, each position—ones, tens, and hundreds places—is a single-digit answer, so the truck may be loaded as you go.)
2. Have students work on a problem involving a triple digit using blackline master S8.3 (page 161). Let them work independently or in pairs.
3. Ask students how they would solve this problem: 3 × 257. Work through the problem with them as you did earlier, only this time, you will need to add a thousands room to your artwork. "Cards" are "posted on the wall" from the room at the far right to the room at the far left, just as they were posted when problems involving two-digit numbers were solved.

4. When working out the problem 3 × 257, have the children write their work like this:

$$
\begin{array}{r}
6 \quad (3 \times 2 = 6) \\
1\ 5 \quad (3 \times 5 = 15) \\
2 \\
257 \\
\times 3 \quad (3 \times 7 = 21) \\
\hline
771 \\
\end{array}
$$

Multiply, post numbers on the wall, then load the truck!

This practice will help both student and teacher see the computation and thus find problem areas.

5. Solve a few more problems with your students, then give them the challenge problems found on blackline master S8.4 (page 162). Of course, let them work in pairs if it will help them.

Assess

Assess students' mastery of the subject using assessments T8.2 and T8.3 (pages 195–196). Have students show all their work. Grade students' papers in order to determine each student's understanding of the process of multi-digit multiplication. Pay attention to the areas in which student understanding seems weak, if any, and reteach the lessons, focusing on those areas, if necessary. Provide students frequent opportunities to practice; practicing for five minutes at the beginning of each math lesson is recommended.

Use challenge assessment T8.4 (page 197) to test students' mastery of three- and four-place multiplication.

Multi-Digit Division

rs. Twig and her bakery once again appear in the story used to illustrate division. Like the story used to illustrate multiplication, this story involves posting numbers, loading trucks, and using place values. This story is set up differently, and the drawing for the story is a little different, but the similarities between the two stories are so numerous that students are not likely to experience this work as being unrelated to the work of multi-digit multiplication.

Introduce

Mrs. Twig's business grew and grew. People came from far and wide to buy some of Mrs. Twig's cookies! On the first day of each month, Mrs. Twig had a special sale that she called "Bakery Blowout." Her bakers worked hard to fill the shelves of the bakery with boxes of cookies, and when the store opened for business on those days, there were always people waiting on the doorstep to come in and buy cookies.

May 1 was a "Bakery Blowout" day. On that day, 5 people had driven their trucks to Mrs. Twig's bakery, parked in her parking lot, and stood on her doorstep, waiting patiently for the doors to open. They knew that each would get an equal number of all of the cookies in the whole bakery. Here is a picture of the "Bakery Blowout" just before Mrs. Twig opened her bakery for business that day.

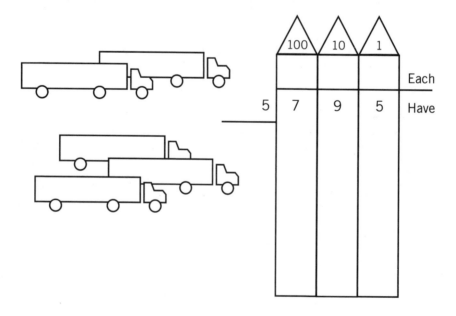

Procedure

Tell students,

1. "Look at the picture. Do you see the 5 trucks parked in the lot? How about their owners standing on the doorstep?"
2. "Do you see the 3 rooms in the bakery?" (hundreds, tens, and ones)
3. "Tell me how many boxes of 100 cookies there are in the bakery today." (7)
4. "How many boxes of 10 cookies are there today?" (9)
5. "How many individual cookies are there?" (5)
6. "Find the word 'have.' This tells us the number of cookies or boxes of cookies that we have available to sell at that moment."
7. "Find the board labeled 'each' that Mrs. Twig uses to write down how many boxes each buyer gets. Mrs. Twig writes her answers on that board as she figures out how many boxes of cookies people need. That way, customers can see right away how many of each kind of box they should buy."
8. "Customers look at the board, then take the correct number of each box of cookies and put them on their trucks right away."
9. "On this particular day, the 5 buyers came into the hundreds room. Mrs. Twig saw at a glance that there were enough boxes to give each buyer one box. She did not have enough boxes to give them each two boxes. Mrs. Twig quickly wrote the number 1 on the board over the hundreds house."

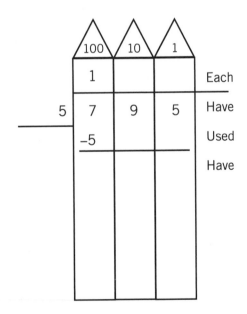

10. "The customers happily took their boxes of 100 cookies out to their trucks."

11. "Meanwhile, Mrs. Twig went back downstairs and figured out the number of cookies she had just sold and how many she had left. 'I have given 1 box 5 times,' she said. 'I used five boxes.' She wrote this in the hundreds room like this: - 5."

12. "Next, Mrs. Twig figured out how many boxes of cookies she had left in each room. There were 2 boxes of 100 cookies (2 hundreds), 9 boxes of 10 cookies (9 tens), and 5 single cookies (5 ones).

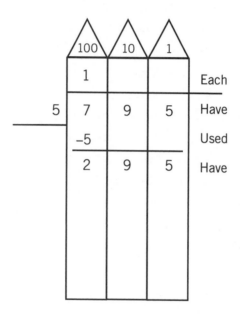

13. "The customers returned from loading their trucks and found Mrs. Twig looking inside the 2 hundred-cookie boxes that she had left. She noted that each box had 10 10-cookie boxes inside of them. Mrs. Twig took the 20 smaller boxes out of their big boxes, laid the 20 boxes of tens on a tray, and carried them with her into the tens room. That is where they belonged now." Draw an arrow under the number 2, in the hundreds room, to show that it has moved to the tens room. Next, draw a circle around the numbers 2 and 9 and say, "She has 29 tens boxes in this room."

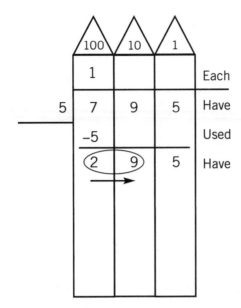

14. "Mrs. Twig began to divide the 29 tens boxes among the 5 buyers. She found that she had enough to give each buyer 5 boxes, but not enough to give each 6. Mrs. Twig ran up the stairs and wrote a big number 5 on the board over the tens room."

15. "The customers immediately took five boxes each and loaded them on their trucks."

16. "Mrs. Twig said, 'I sold 5 boxes 5 times. This means I sold 25 boxes in all.' She wrote the number 25 under the 29 that was there. Then she figured out how many boxes she had left to sell."

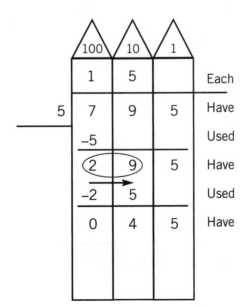

17. "Mrs. Twig saw that she had 0 boxes in the hundreds room, 4 boxes in the tens room, and 5 cookies in the ones room."

18. "The buyers returned, and Mrs. Twig opened the four tens boxes to find that each box had 10 cookies in it. She carefully put the 40 cookies on a tray and carried them into the ones room." Draw an arrow showing that the number 4 went next door. Circle the numbers 4 and 5 to show that now there are 45 individual cookies, or ones.

19. "Mrs. Twig divided 45 by 5 and discovered that she could give each customer 9 cookies. She went upstairs and wrote the number 9 on the board."

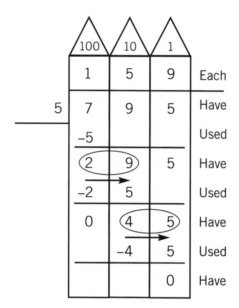

Art 9.6

20. "The buyers each happily gathered their 9 cookies and took them out to their trucks."

21. "Mrs. Twig went back downstairs to calculate how many cookies she'd sold altogether and how many were left. This is what she came up with: each customer got 159 cookies—there was nothing left!"

Practice

Oral Practice

Do several more problems together, with students supplying answers. Go slowly, and be sure that you follow the same steps each time. There is a rhythm to working long division; that rhythm helps students understand what is going on.

Written Practice

When students seem to understand how to move through a problem correctly, give them each a copy of blackline master S9.1 (page 163) and have them solve the first problem while still seated near you. Discuss with them how it went, answer their questions, and move on to the second problem on the sheet. Again, discuss their experiences in solving the problem and answer their questions. If work is progressing smoothly, let students work on the rest of the problems on their own. Take this opportunity to help any struggling students, but do not reteach the lesson; instead, guide them through the story by asking questions, such as "What did Mrs. Twig do next?" You might ask them to stop frequently to visualize the action of the story. You might even ask them what color the rooms are painted. Can they see Mrs. Twig and the buyers?

Hints

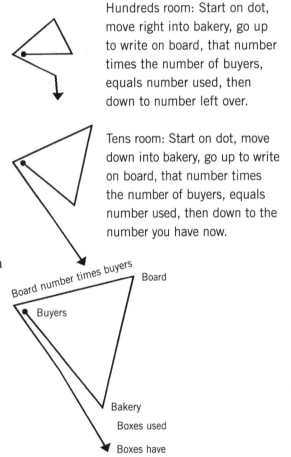

- A kinesthetic tie to the action in the story is that the movement flows in ever-widening triangle shapes. This is what the triangle shapes would look like, if you were to graph the action as though someone were walking through the graph with paint on the bottom of his or her shoes.

Hundreds room: Start on dot, move right into bakery, go up to write on board, that number times the number of buyers, equals number used, then down to number left over.

Tens room: Start on dot, move down into bakery, go up to write on board, that number times the number of buyers, equals number used, then down to the number you have now.

- Pair up students who work well together without losing their focus and without one child dominating the other. Have them take turns "teaching" the action of the story to each other. This practice facilitates fluency, and there is magic in the act of teaching a concept to another person.

- Instead of having students raise their hands (and possibly admit confusion publicly), conduct a secret ballot at this point in the lesson. Give students scrap paper and have them put their names on their papers. Then ask them to answer the following questions:

1. Are you a one, two, three, or four? One = I get it! Two = I pretty much have it, but I need more practice. Three = I'm confused, although I get some of it. Four = I'm totally lost!
2. If you are a one, what helped you? If you are a two, what part is hard for you? If you are a three, what part confuses you? If you are a four, how far did you get before you got lost?
3. Any other comments?

Collect the secret surveys and read them. Remove the "ones," and be happy. Group the "twos" together and check for areas in which they need practice. Pair these students up to practice together. Group the "threes" together and check for common problem areas. Sometimes these students need to review the procedures involved. Sometimes they are shaky on the logic behind the story. At any rate, judiciously pair each "three" with a "one." Have the "ones" teach the story to the "threes," then have the "threes" teach it to the "ones" as many times as they need to practice.

Collect the "fours" and go over the lesson again. Do not use the same language that you used the first time. It did not do the job, so now you need to adjust your method. Ask students to find the doorstep, the board, and the bakery rooms (hundreds, tens, and ones). Have them find and tell you how many cookies are available to buy. What does the number on the doorstep represent? What do the numbers in the attics represent? What are the buyers doing at the bakery?

Ask next, "What do we do first?" Have a buyer knock on the door to begin the action. Ask questions at every step in order to engage the children's logic and sense of story. Draw on their ability to envision a scene. (Example: "I see Mrs. Twig writing a number on the board. The wind is blowing hard and her apron is whipping around her head! The buyers are standing on the ground watching her write . . . there they go! They are getting their boxes and running out to their trucks!")

✎ Notice also that the actors take turns in the action:
1. The buyers come in.
2. Mrs. Twig figures out how many boxes each buyer gets and writes the number on her board.

3. The buyers load their cookies on trucks.
4. Mrs. Twig figures out how many boxes she sold and how many she has now.
5. Action repeats.

Give students extra practice using copies of blackline masters S9.2, S9.3, and S9.4 (pages 164–166). Notice that blackline master S9.4 involves remainders. Simply explain the remainder as the number that is left over: "what remains."

Challenge

Blackline masters S9.5 and S9.6 (pages 167–168) provide even more challenge for those students who want it. These challenges involve dividends that have more than three digits. They also challenge students to handle long division without looking at pictures. (For a four-digit dividend, simply add a thousands room.)

Assess

Assess students' mastery using blackline masters S9.7 and S9.8 (pages 169–170). Have students show all their work. Grade the assessments and determine each student's understanding of the process of multi-digit division. Pay attention to the areas in which student understanding seems weak, if any, and reteach the lessons, focusing on those areas. Provide students frequent opportunities to practice, even five minutes a day as a warm-up activity.

Appendix A

Student Blackline Masters

S1.1

1	2	3	4	5
6	7	8	9	10
11	12	13	14	15
16	17	18	19	20

S1.2

1	2	3	4	5
6	7	8	9	10
11	12	13	14	15

S1.3

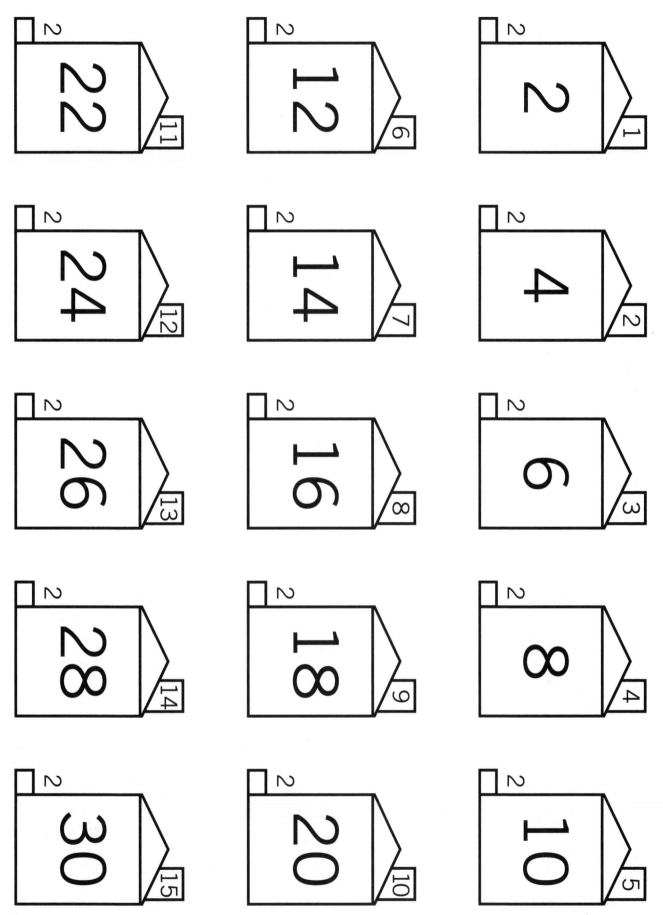

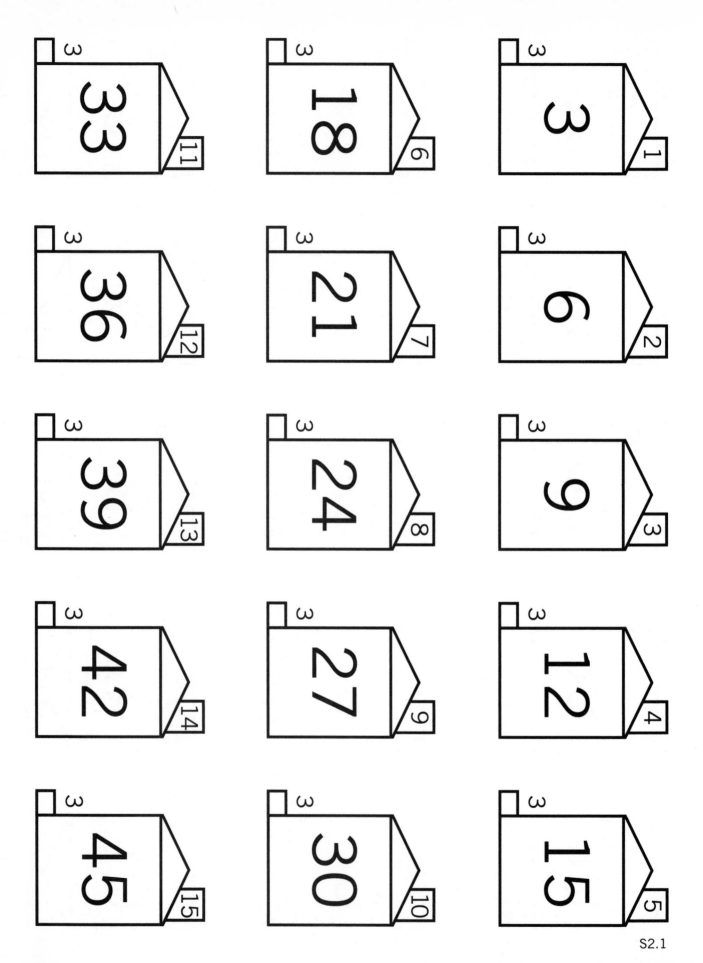

S2.1

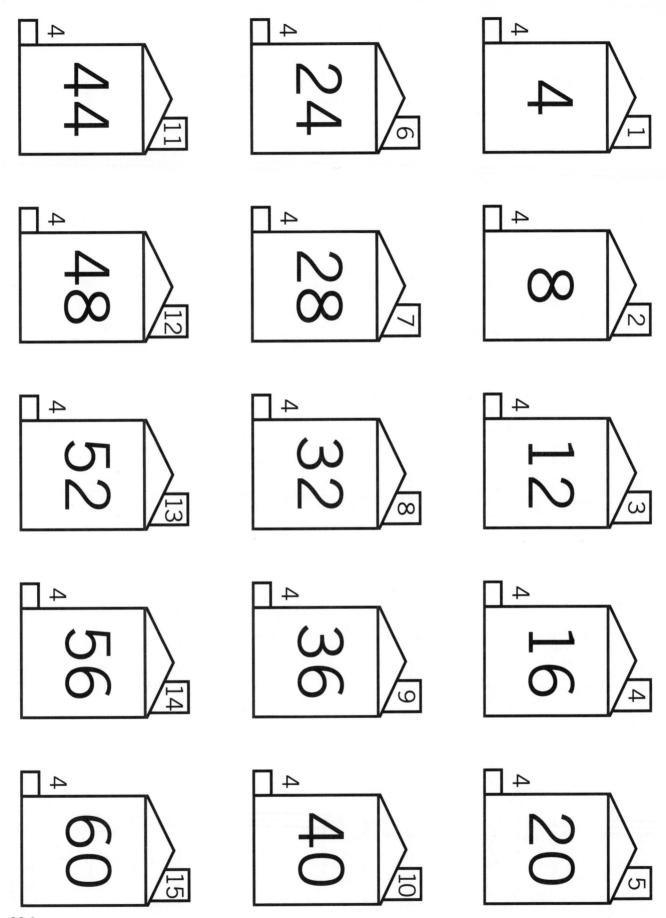

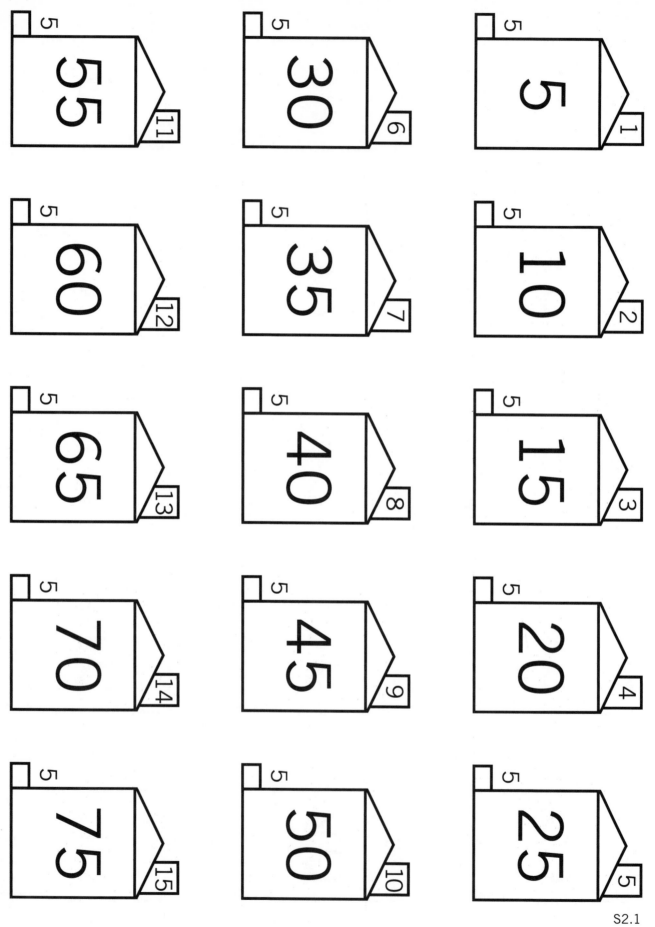

S2.1

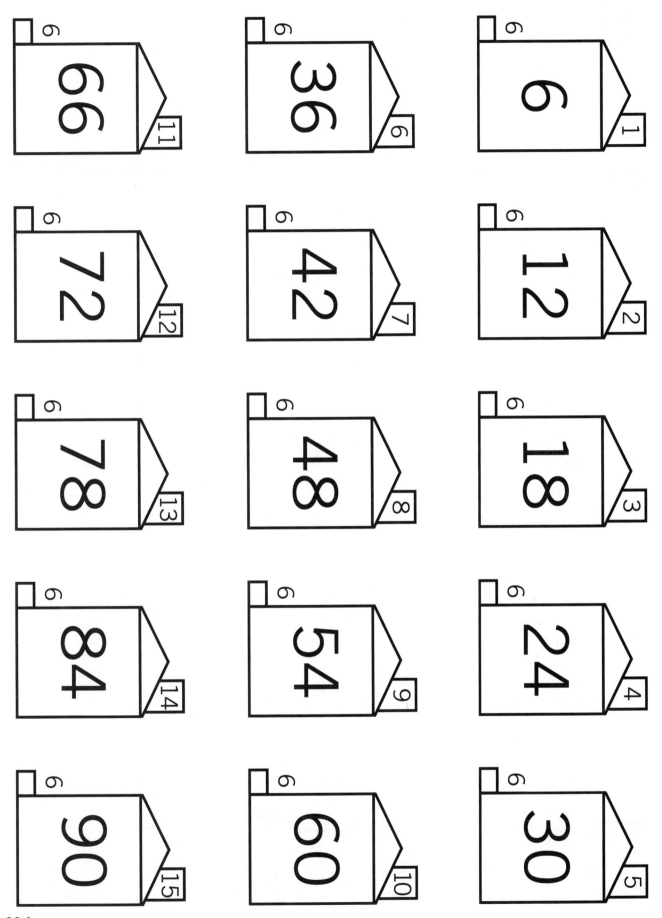

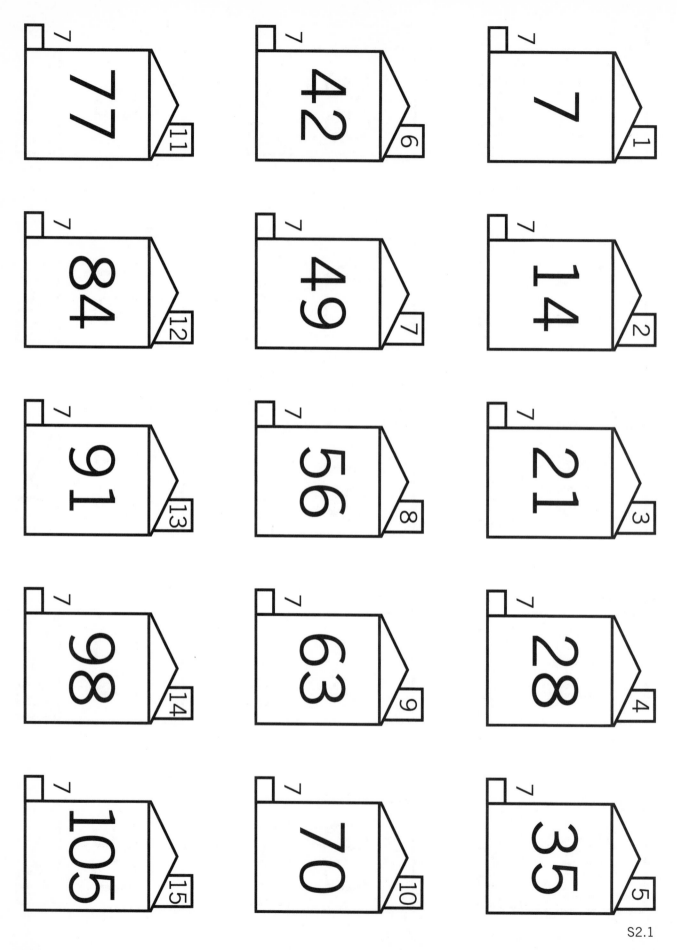

S2.1

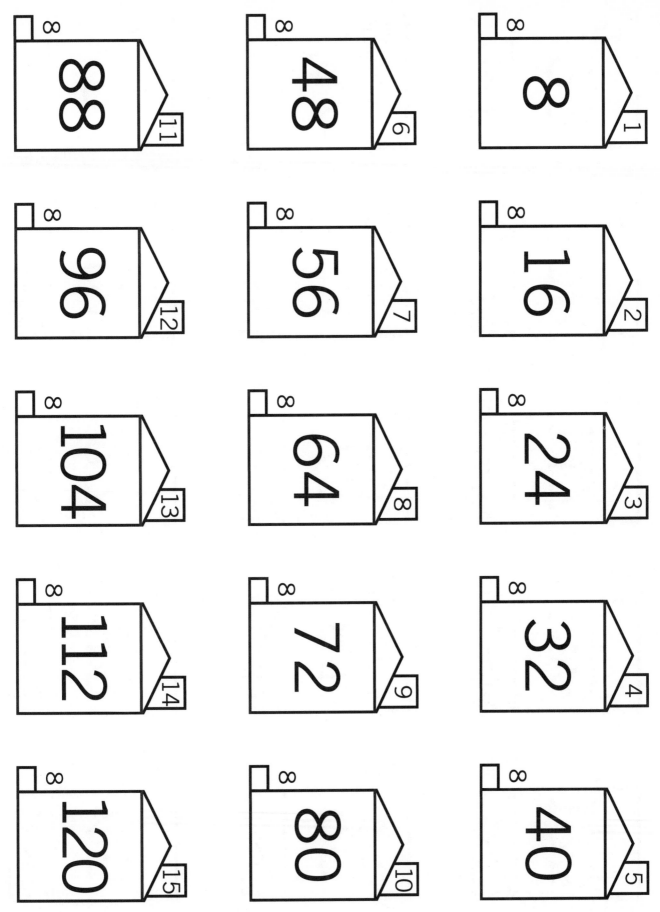

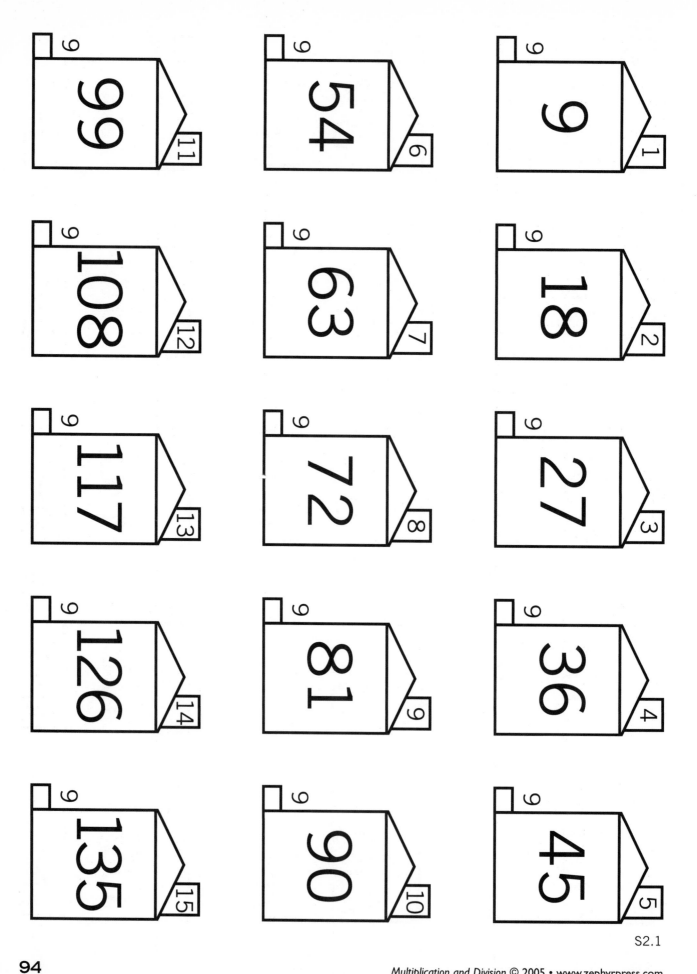

S2.1

94

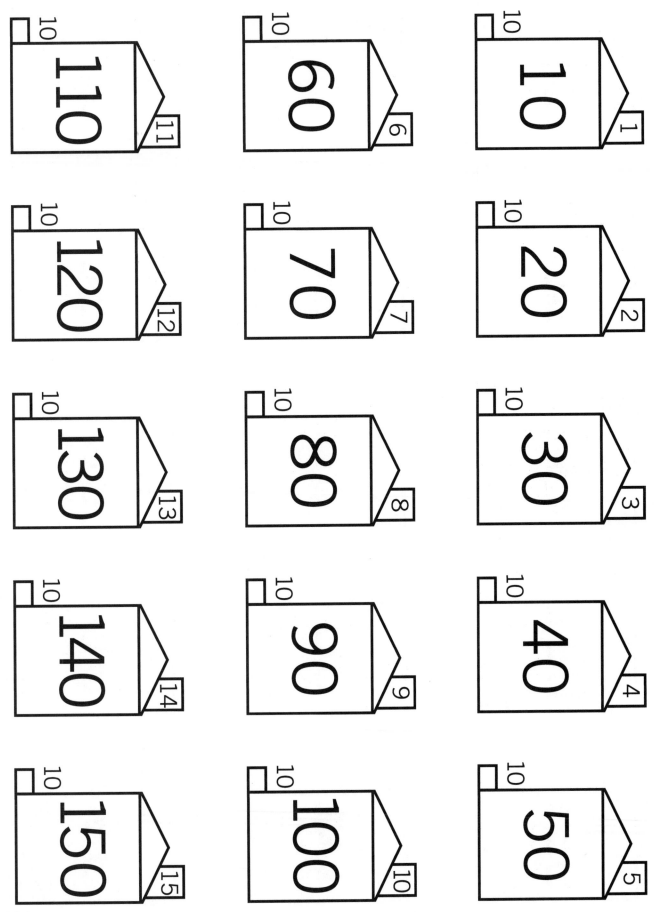

S2.1

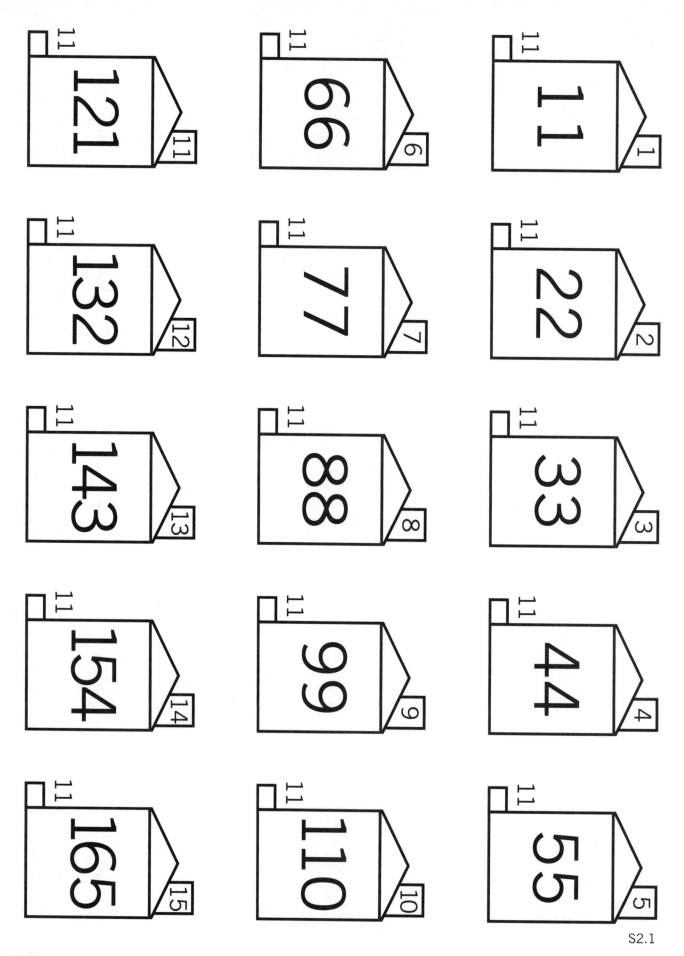

S2.1

Multiplication and Division © 2005 • www.zephyrpress.com

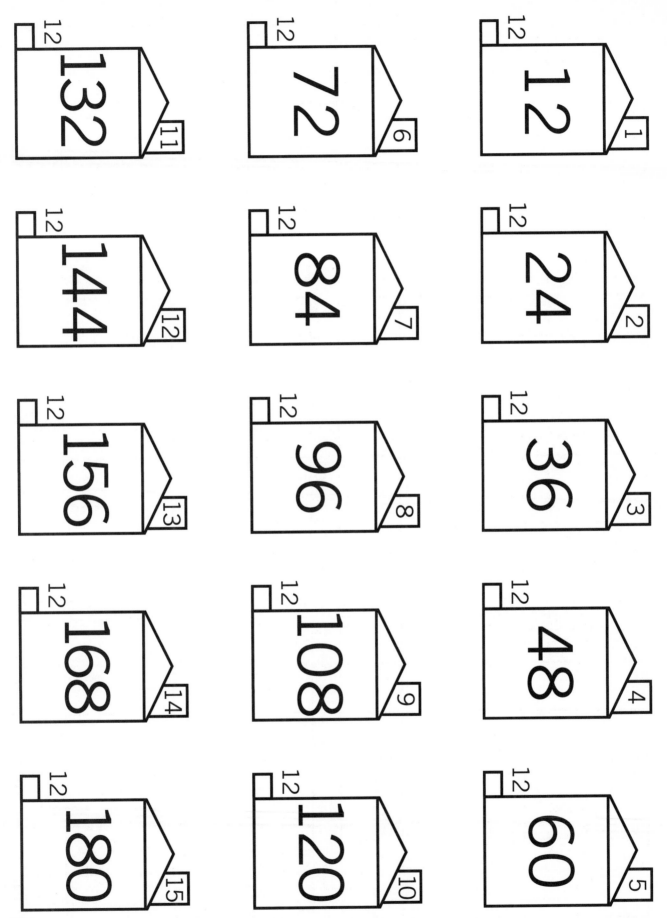

S2.1

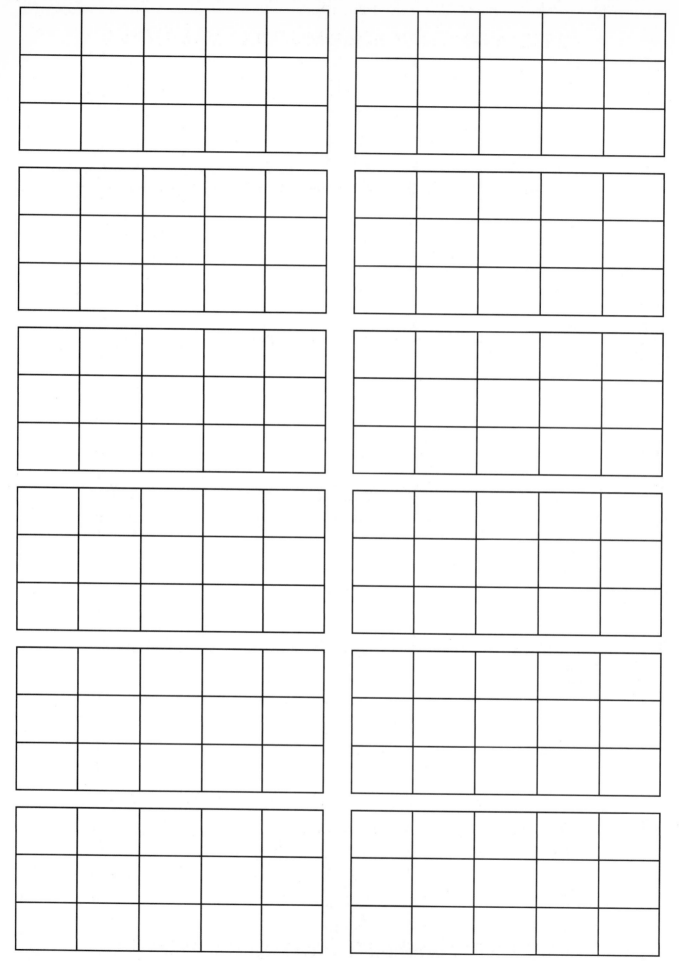

S3.1

Multiplication and Division © 2005 • www.zephyrpress.com

Master Map

1	2	3	4	5
6	7	8	9	10
11	12	13	14	15

2	4	6	8	10
12	14	16	18	20
22	24	26	28	30

3	6	9	12	15
18	21	24	27	30
33	36	39	42	45

4	8	12	16	20
24	28	32	36	40
44	48	52	56	60

5	10	15	20	25
30	35	40	45	50
55	60	65	70	75

6	12	18	24	30
36	42	48	54	60
66	72	78	84	90

7	14	21	28	35
42	49	56	63	70
77	84	91	98	105

8	16	24	32	40
48	56	64	72	80
88	96	104	112	120

9	18	27	36	45
54	63	72	81	90
99	108	117	126	135

10	20	30	40	50
60	70	80	90	100
110	120	130	140	150

11	22	33	44	55
66	77	88	99	110
121	132	143	154	165

12	24	36	48	60
72	84	96	108	120
132	144	156	168	180

S3.2

Multiplication and Division © 2005 • www.zephyrpress.com

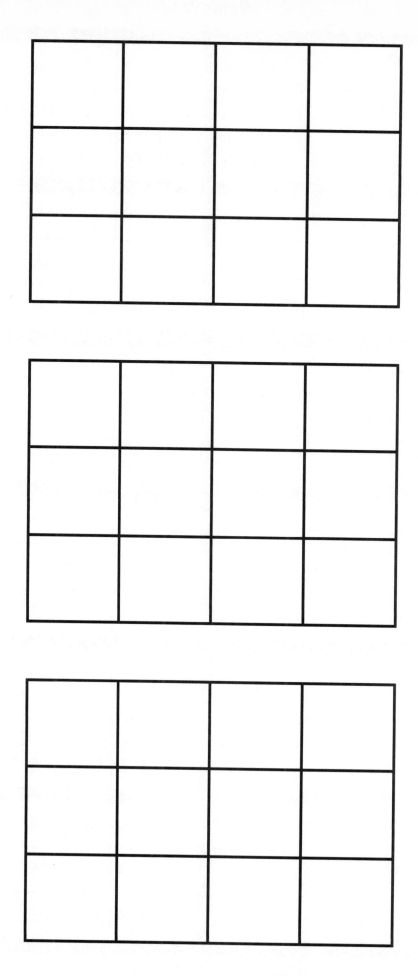

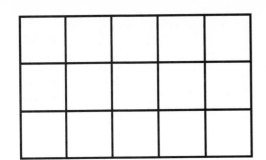

Name: _____

$$\begin{array}{r} 5 \\ \times 10 \\ \hline \end{array}$$
$$\begin{array}{r} 8 \\ \times 10 \\ \hline \end{array}$$
$$\begin{array}{r} 7 \\ \times 10 \\ \hline \end{array}$$
$$\begin{array}{r} 11 \\ \times 10 \\ \hline \end{array}$$
$$\begin{array}{r} 6 \\ \times 10 \\ \hline \end{array}$$

$$\begin{array}{r} 9 \\ \times 10 \\ \hline \end{array}$$
$$\begin{array}{r} 12 \\ \times 10 \\ \hline \end{array}$$
$$\begin{array}{r} 10 \\ \times 10 \\ \hline \end{array}$$
$$\begin{array}{r} 3 \\ \times 10 \\ \hline \end{array}$$
$$\begin{array}{r} 9 \\ \times 10 \\ \hline \end{array}$$

$$\begin{array}{r} 2 \\ \times 10 \\ \hline \end{array}$$
$$\begin{array}{r} 1 \\ \times 10 \\ \hline \end{array}$$
$$\begin{array}{r} 7 \\ \times 10 \\ \hline \end{array}$$
$$\begin{array}{r} 13 \\ \times 10 \\ \hline \end{array}$$
$$\begin{array}{r} 5 \\ \times 10 \\ \hline \end{array}$$

$$\begin{array}{r} 8 \\ \times 10 \\ \hline \end{array}$$
$$\begin{array}{r} 11 \\ \times 10 \\ \hline \end{array}$$
$$\begin{array}{r} 4 \\ \times 10 \\ \hline \end{array}$$
$$\begin{array}{r} 14 \\ \times 10 \\ \hline \end{array}$$
$$\begin{array}{r} 15 \\ \times 10 \\ \hline \end{array}$$

Tens

The Multiplication and Division Connection

Fingertip Animal Cards

Division

One day we made some cards in art class. Each of us dipped all ten fingertips in ink and made fingerprints on cards with each finger. When the cards dried, we decorated them with a felt-tip pen, making eyes, legs, and things like that. Of course, each of us made 10 cards.

Choose a TN of cards made in all. Divide by number of fingers on each child.

How many children made fingertip animal cards?

Procedure: TN ÷ number of cards

Target numbers: answers on houses chart

From Real to Symbolic:
Use pieces of small tagboard for the children to create and solve the problems.

Next, use plastic chips or some other tangible objects to represent the cards.

Last, use number symbols.

$$10\overline{)60}$$

6 number of children making cards

10 fingers

60 total number of animal cards

Multiplication

Choose a number of children that want to work on making fingerprint animal cards.

How many fingerprint animal cards would you end up with when your project is finished?

Procedure: number of fingers X TN (Start with two smaller numbers and work with them to arrive at large number.)

Target numbers to choose from:
1–15

S7.2

102

Multiplication and Division © 2005 • www.zephyrpress.com

Name: _____

$10\overline{)90}$ $\quad$ $10\overline{)100}$ $\quad$ $10\overline{)70}$ $\quad$ $10\overline{)20}$ $\quad$ $10\overline{)60}$

$10\overline{)110}$ $\quad$ $10\overline{)140}$ $\quad$ $10\overline{)150}$ $\quad$ $10\overline{)130}$ $\quad$ $10\overline{)90}$

$10\overline{)120}$ $\quad$ $10\overline{)140}$ $\quad$ $10\overline{)50}$ $\quad$ $10\overline{)80}$ $\quad$ $10\overline{)40}$

$10\overline{)30}$ $\quad$ $10\overline{)100}$ $\quad$ $10\overline{)150}$ $\quad$ $10\overline{)70}$ $\quad$ $10\overline{)110}$

S7.3

Multiplication and Division © 2005 • www.zephyrpress.com

$$
\begin{array}{r} 5 \\ \times 2 \\ \hline \end{array}
\qquad
\begin{array}{r} 8 \\ \times 2 \\ \hline \end{array}
\qquad
\begin{array}{r} 7 \\ \times 2 \\ \hline \end{array}
\qquad
\begin{array}{r} 11 \\ \times 2 \\ \hline \end{array}
\qquad
\begin{array}{r} 6 \\ \times 2 \\ \hline \end{array}
$$

$$
\begin{array}{r} 9 \\ \times 2 \\ \hline \end{array}
\qquad
\begin{array}{r} 12 \\ \times 2 \\ \hline \end{array}
\qquad
\begin{array}{r} 10 \\ \times 2 \\ \hline \end{array}
\qquad
\begin{array}{r} 3 \\ \times 2 \\ \hline \end{array}
\qquad
\begin{array}{r} 14 \\ \times 2 \\ \hline \end{array}
$$

$$
\begin{array}{r} 2 \\ \times 2 \\ \hline \end{array}
\qquad
\begin{array}{r} 1 \\ \times 2 \\ \hline \end{array}
\qquad
\begin{array}{r} 7 \\ \times 2 \\ \hline \end{array}
\qquad
\begin{array}{r} 13 \\ \times 2 \\ \hline \end{array}
\qquad
\begin{array}{r} 5 \\ \times 2 \\ \hline \end{array}
$$

$$
\begin{array}{r} 8 \\ \times 2 \\ \hline \end{array}
\qquad
\begin{array}{r} 11 \\ \times 2 \\ \hline \end{array}
\qquad
\begin{array}{r} 4 \\ \times 2 \\ \hline \end{array}
\qquad
\begin{array}{r} 14 \\ \times 2 \\ \hline \end{array}
\qquad
\begin{array}{r} 15 \\ \times 2 \\ \hline \end{array}
$$

S7.4

Twos

The Multiplication and Division Connection

Division

Jack and Jill hunt for bugs together. After their hunt, they divide their bugs into two jars so that they each have the same number of bugs.

Choose a target number (TN) of bugs they found from the twos chart and put that number in the large jar.

Give each child the same number of bugs. How many do they each get?

Procedure: TN ÷ number of kids

Target numbers: answers on houses chart

From Real to Symbolic

Use plastic bugs for the children to create and solve the problems.

Next, use plastic chips or some other tangible objects.

Last, use number symbols.

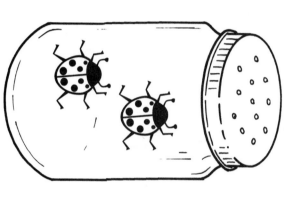

Our Bugs

Jack's Bugs

Jill's Bugs

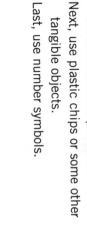

22 number for each child

44 total number of bugs found

2 kids

Multiplication

Jack and Jill go on separate hunts in order to cover more territory. They each have their own jar to put bugs in.

Choose a target number of bugs they each found.

When they were finished with their hunt, they put all their bugs together in the large jar.

How many bugs did they have altogether?

Procedure: number of kids X TN (Start with two smaller numbers and work them to arrive at large number.)

Target numbers to choose from: 1–15

S7.5

105

Multiplication and Division © 2005 • www.zephyrpress.com

$2\overline{)18}$ $2\overline{)8}$ $2\overline{)12}$ $2\overline{)20}$ $2\overline{)14}$

$2\overline{)10}$ $2\overline{)24}$ $2\overline{)6}$ $2\overline{)22}$ $2\overline{)30}$

$2\overline{)30}$ $2\overline{)16}$ $2\overline{)2}$ $2\overline{)4}$ $2\overline{)20}$

$2\overline{)10}$ $2\overline{)8}$ $2\overline{)26}$ $2\overline{)14}$ $2\overline{)28}$

S7.6

Multiplication and Division © 2005 • www.zephyrpress.com

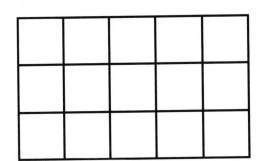

Name: _____

2 ×10	8 ×2	7 ×10	6 ×10	6 ×2
9 ×2	12 ×10	10 ×2	3 ×10	14 ×2
9 ×10	11 ×2	13 ×2	14 ×10	5 ×2
7 ×2	8 ×10	4 ×10	15 ×2	5 ×10
11 ×10	12 ×2	10 ×10	15 ×10	13 ×10

S7.7

$2\overline{)18}$ $10\overline{)100}$ $2\overline{)8}$ $10\overline{)60}$ $2\overline{)30}$

$2\overline{)20}$ $10\overline{)70}$ $2\overline{)10}$ $2\overline{)30}$ $2\overline{)24}$

$2\overline{)110}$ $10\overline{)140}$ $10\overline{)150}$ $10\overline{)130}$ $10\overline{)90}$

$2\overline{)14}$ $2\overline{)20}$ $2\overline{)28}$ $2\overline{)26}$ $2\overline{)22}$

$2\overline{)4}$ $2\overline{)8}$ $10\overline{)30}$ $2\overline{)6}$ $10\overline{)40}$

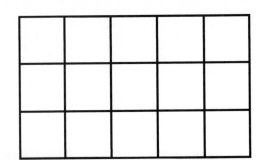

5 ×12	8 ×12	7 ×12	11 ×12	6 ×12
9 ×12	12 ×12	10 ×12	3 ×12	9 ×12
2 ×12	1 ×12	7 ×12	13 ×12	5 ×12
8 ×12	11 ×12	4 ×12	14 ×12	15 ×12

Twelves

Division

We are going to boil eggs to decorate for our Easter baskets. We have (TN) eggs in a big bowl! There are so many! There are 12 children who are making baskets.

How many eggs will we each get to decorate?

Procedure: TN ÷ number of children

Target numbers: answers on houses chart

From Real to Symbolic:

Draw pictures of baskets or cut out pictures of children for the students to create and solve the problems.

Next, use plastic chips or some other tangible objects to represent the baskets or kids.

Last, use number symbols.

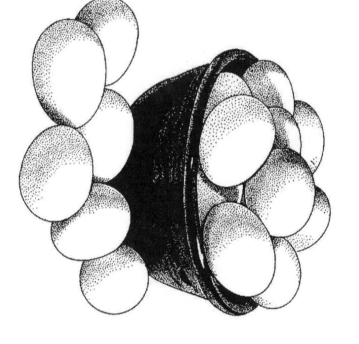

Our Easter Eggs

Multiplication

We made baskets for our Easter eggs in art class. Each of our baskets will hold TN eggs.

How many eggs will we need to buy in order for each of us to have that many eggs?

Procedure: number of kids X TN (Start with two smaller numbers and work them to arrive at large number.)

Target numbers to choose from: 1–15

$$12\overline{)72}$$

6 number of eggs per child

12 children

72 total number of eggs decorated

S7.9

$12\overline{)96}$ $12\overline{)120}$ $12\overline{)60}$ $12\overline{)72}$ $12\overline{)24}$

$12\overline{)48}$ $12\overline{)168}$ $12\overline{)108}$ $12\overline{)144}$ $12\overline{)156}$

$12\overline{)132}$ $12\overline{)72}$ $12\overline{)96}$ $12\overline{)120}$ $12\overline{)84}$

$12\overline{)108}$ $12\overline{)84}$ $12\overline{)180}$ $12\overline{)132}$ $12\overline{)24}$

S7.10

$$\begin{array}{r}4\\ \times12\\ \hline\end{array} \qquad \begin{array}{r}8\\ \times12\\ \hline\end{array} \qquad \begin{array}{r}2\\ \times12\\ \hline\end{array} \qquad \begin{array}{r}7\\ \times10\\ \hline\end{array} \qquad \begin{array}{r}13\\ \times12\\ \hline\end{array}$$

$$\begin{array}{r}9\\ \times12\\ \hline\end{array} \qquad \begin{array}{r}12\\ \times10\\ \hline\end{array} \qquad \begin{array}{r}9\\ \times12\\ \hline\end{array} \qquad \begin{array}{r}7\\ \times2\\ \hline\end{array} \qquad \begin{array}{r}3\\ \times10\\ \hline\end{array}$$

$$\begin{array}{r}9\\ \times10\\ \hline\end{array} \qquad \begin{array}{r}11\\ \times2\\ \hline\end{array} \qquad \begin{array}{r}14\\ \times10\\ \hline\end{array} \qquad \begin{array}{r}13\\ \times2\\ \hline\end{array} \qquad \begin{array}{r}12\\ \times12\\ \hline\end{array}$$

$$\begin{array}{r}6\\ \times12\\ \hline\end{array} \qquad \begin{array}{r}8\\ \times10\\ \hline\end{array} \qquad \begin{array}{r}3\\ \times12\\ \hline\end{array} \qquad \begin{array}{r}10\\ \times12\\ \hline\end{array} \qquad \begin{array}{r}4\\ \times10\\ \hline\end{array}$$

S7.11

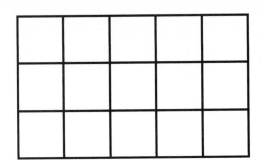

$$\begin{array}{r} 7 \\ \times 12 \\ \hline \end{array}$$ $$\begin{array}{r} 10 \\ \times 10 \\ \hline \end{array}$$ $$\begin{array}{r} 2 \\ \times 2 \\ \hline \end{array}$$ $$\begin{array}{r} 12 \\ \times 2 \\ \hline \end{array}$$ $$\begin{array}{r} 11 \\ \times 10 \\ \hline \end{array}$$

$$\begin{array}{r} 4 \\ \times 2 \\ \hline \end{array}$$ $$\begin{array}{r} 6 \\ \times 10 \\ \hline \end{array}$$ $$\begin{array}{r} 6 \\ \times 2 \\ \hline \end{array}$$ $$\begin{array}{r} 5 \\ \times 12 \\ \hline \end{array}$$ $$\begin{array}{r} 3 \\ \times 2 \\ \hline \end{array}$$

$$\begin{array}{r} 5 \\ \times 10 \\ \hline \end{array}$$ $$\begin{array}{r} 8 \\ \times 12 \\ \hline \end{array}$$ $$\begin{array}{r} 5 \\ \times 2 \\ \hline \end{array}$$ $$\begin{array}{r} 14 \\ \times 2 \\ \hline \end{array}$$ $$\begin{array}{r} 15 \\ \times 12 \\ \hline \end{array}$$

$$\begin{array}{r} 15 \\ \times 10 \\ \hline \end{array}$$ $$\begin{array}{r} 14 \\ \times 12 \\ \hline \end{array}$$ $$\begin{array}{r} 11 \\ \times 12 \\ \hline \end{array}$$ $$\begin{array}{r} 15 \\ \times 2 \\ \hline \end{array}$$ $$\begin{array}{r} 13 \\ \times 10 \\ \hline \end{array}$$

S7.11

$$2\overline{)16} \qquad 10\overline{)90} \qquad 12\overline{)84} \qquad 10\overline{)50} \qquad 12\overline{)60}$$

$$2\overline{)8} \qquad 10\overline{)140} \qquad 2\overline{)12} \qquad 12\overline{)96} \qquad 2\overline{)24}$$

$$10\overline{)130} \qquad 12\overline{)108} \qquad 12\overline{)144} \qquad 2\overline{)30} \qquad 12\overline{)132}$$

$$2\overline{)14} \qquad 12\overline{)24} \qquad 2\overline{)28} \qquad 12\overline{)36} \qquad 2\overline{)18}$$

$$2\overline{)4} \qquad 2\overline{)26} \qquad 10\overline{)30} \qquad 12\overline{)72} \qquad 10\overline{)40}$$

S7.11

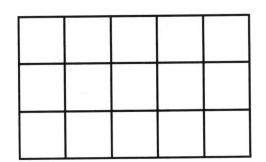

$$
\begin{array}{r} 5 \\ \times 8 \\ \hline \end{array}
\qquad
\begin{array}{r} 8 \\ \times 8 \\ \hline \end{array}
\qquad
\begin{array}{r} 7 \\ \times 8 \\ \hline \end{array}
\qquad
\begin{array}{r} 11 \\ \times 8 \\ \hline \end{array}
\qquad
\begin{array}{r} 6 \\ \times 8 \\ \hline \end{array}
$$

$$
\begin{array}{r} 9 \\ \times 8 \\ \hline \end{array}
\qquad
\begin{array}{r} 12 \\ \times 8 \\ \hline \end{array}
\qquad
\begin{array}{r} 10 \\ \times 8 \\ \hline \end{array}
\qquad
\begin{array}{r} 3 \\ \times 8 \\ \hline \end{array}
\qquad
\begin{array}{r} 9 \\ \times 8 \\ \hline \end{array}
$$

$$
\begin{array}{r} 2 \\ \times 8 \\ \hline \end{array}
\qquad
\begin{array}{r} 1 \\ \times 8 \\ \hline \end{array}
\qquad
\begin{array}{r} 7 \\ \times 8 \\ \hline \end{array}
\qquad
\begin{array}{r} 13 \\ \times 8 \\ \hline \end{array}
\qquad
\begin{array}{r} 5 \\ \times 8 \\ \hline \end{array}
$$

$$
\begin{array}{r} 8 \\ \times 8 \\ \hline \end{array}
\qquad
\begin{array}{r} 11 \\ \times 8 \\ \hline \end{array}
\qquad
\begin{array}{r} 4 \\ \times 8 \\ \hline \end{array}
\qquad
\begin{array}{r} 14 \\ \times 8 \\ \hline \end{array}
\qquad
\begin{array}{r} 15 \\ \times 8 \\ \hline \end{array}
$$

Eights

Division

We are making clay spiders in class this week. We are using air-drying clay for the bodies and beads for the eyes. For the legs, we are cutting pipe cleaners. We have cut a huge pile of legs (choose TN), and we need to put eight on each body.

How many spiders can we make with the legs we've cut?

Target numbers: answers on houses chart

Procedure: TN ÷ number of legs

From Real to Symbolic:

Use actual materials (pipe cleaner legs, etc.) for the children to create and solve the problems.

Next, use plastic chips or some other tangible objects to represent the spiders.

Last, use number symbols.

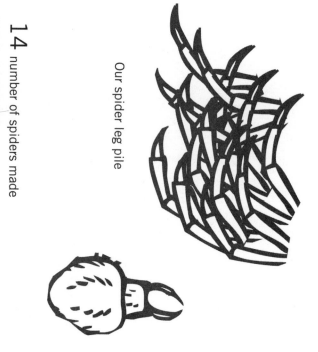

Our spider leg pile

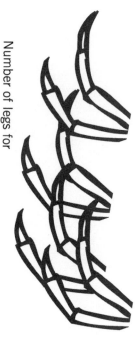

Number of legs for each spider

Multiplication

Choose a number of spiders you want to make. You know you need to cut 8 legs for each spider.

How many legs do you need to cut altogether?

Procedure: number of legs X TN (Start with two smaller numbers and work them to arrive at large number.)

Target numbers to choose from:
1–15

$$\begin{array}{r} 14 \\ 8\overline{)112} \end{array}$$

14 number of spiders made

112 total number of legs cut

8 legs

S7.13

Multiplication and Division © 2005 • www.zephyrpress.com

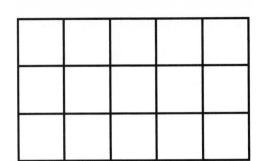

Name: _____

$8\overline{)16}$　　$8\overline{)72}$　　$8\overline{)120}$　　$8\overline{)24}$　　$8\overline{)64}$

$8\overline{)88}$　　$8\overline{)56}$　　$8\overline{)32}$　　$8\overline{)48}$　　$8\overline{)104}$

$8\overline{)64}$　　$8\overline{)72}$　　$8\overline{)112}$　　$8\overline{)40}$　　$8\overline{)80}$

$8\overline{)96}$　　$8\overline{)48}$　　$8\overline{)16}$　　$8\overline{)56}$　　$8\overline{)40}$

S7.14

4	2	2	10	11
×12	×8	×12	×12	×8

9	13	3	9	15
×8	×8	×12	×12	×8

10	12	12	14	13
×8	×2	×8	×10	×2

8	8	6	11	7
×12	×8	×12	×2	×12

S7.15

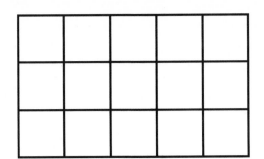

Name: _____

$$\begin{array}{r} 15 \\ \times 12 \\ \hline \end{array}$$
$$\begin{array}{r} 5 \\ \times 12 \\ \hline \end{array}$$
$$\begin{array}{r} 10 \\ \times 12 \\ \hline \end{array}$$
$$\begin{array}{r} 6 \\ \times 2 \\ \hline \end{array}$$
$$\begin{array}{r} 5 \\ \times 8 \\ \hline \end{array}$$

$$\begin{array}{r} 3 \\ \times 8 \\ \hline \end{array}$$
$$\begin{array}{r} 13 \\ \times 12 \\ \hline \end{array}$$
$$\begin{array}{r} 14 \\ \times 2 \\ \hline \end{array}$$
$$\begin{array}{r} 14 \\ \times 12 \\ \hline \end{array}$$
$$\begin{array}{r} 6 \\ \times 8 \\ \hline \end{array}$$

$$\begin{array}{r} 4 \\ \times 8 \\ \hline \end{array}$$
$$\begin{array}{r} 15 \\ \times 10 \\ \hline \end{array}$$
$$\begin{array}{r} 12 \\ \times 12 \\ \hline \end{array}$$
$$\begin{array}{r} 12 \\ \times 10 \\ \hline \end{array}$$
$$\begin{array}{r} 13 \\ \times 10 \\ \hline \end{array}$$

$$\begin{array}{r} 14 \\ \times 8 \\ \hline \end{array}$$
$$\begin{array}{r} 15 \\ \times 2 \\ \hline \end{array}$$
$$\begin{array}{r} 7 \\ \times 8 \\ \hline \end{array}$$
$$\begin{array}{r} 10 \\ \times 10 \\ \hline \end{array}$$
$$\begin{array}{r} 11 \\ \times 12 \\ \hline \end{array}$$

S7.15

$8\overline{)16}$ $10\overline{)90}$ $12\overline{)84}$ $10\overline{)50}$ $12\overline{)60}$

$8\overline{)8}$ $10\overline{)140}$ $8\overline{)24}$ $12\overline{)96}$ $8\overline{)32}$

$10\overline{)130}$ $12\overline{)108}$ $12\overline{)144}$ $2\overline{)30}$ $12\overline{)132}$

$8\overline{)40}$ $12\overline{)24}$ $2\overline{)28}$ $12\overline{)36}$ $2\overline{)18}$

$8\overline{)56}$ $8\overline{)96}$ $2\overline{)24}$ $12\overline{)72}$ $10\overline{)40}$

S7.15

$10\overline{)40}$ $8\overline{)24}$ $12\overline{)84}$ $8\overline{)72}$ $12\overline{)60}$

$2\overline{)8}$ $8\overline{)16}$ $8\overline{)32}$ $12\overline{)96}$ $10\overline{)60}$

$8\overline{)88}$ $12\overline{)108}$ $8\overline{)112}$ $8\overline{)56}$ $12\overline{)132}$

$8\overline{)96}$ $12\overline{)48}$ $2\overline{)30}$ $12\overline{)36}$ $8\overline{)64}$

$2\overline{)36}$ $10\overline{)120}$ $12\overline{)84}$ $8\overline{)104}$ $10\overline{)140}$

S7.15

5		8		7		11		6
×4		×4		×4		×4		×4

9		12		10		3		9
×4		×4		×4		×4		×4

2		1		7		13		5
×4		×4		×4		×4		×4

8		11		4		14		15
×4		×4		×4		×4		×4

Multiplication and Division © 2005 • www.zephyrpress.com

Fours

Division

There are four tracks on which to race matchbox cars.

Each race sees a different number of cars participating.

Choose a TN of total cars racing today. Put that number on the garage. Now, divide evenly into four tracks.

How many cars are on each track?

Procedure: TN ÷ number of tracks

Target numbers: answers on houses chart

From Real to Symbolic:
Use real matchbox cars for the children to create and solve the problems. Next, use plastic chips or some other tangible objects to represent the cars. Last, use number symbols.

$$4\overline{)24}\ {\small\text{total number of cars}}$$

6 number on each track

tracks

Multiplication

Choose a number of cars you would like to see racing on each track today. Have the race and may the best car win.

After the race, all the cars drive to the garage to get checked over and cleaned up.

How many cars are there altogether?

Procedure: number of tracks X TN (Start with two smaller numbers and work them to arrive at large number.)

Target numbers to choose from: 1–15

Multiplication and Division © 2005 • www.zephyrpress.com

$4\overline{)32}$ $4\overline{)8}$ $4\overline{)20}$ $4\overline{)16}$ $4\overline{)12}$

$4\overline{)52}$ $4\overline{)24}$ $4\overline{)28}$ $4\overline{)36}$ $4\overline{)40}$

$4\overline{)44}$ $4\overline{)56}$ $4\overline{)16}$ $4\overline{)48}$ $4\overline{)20}$

$4\overline{)60}$ $4\overline{)24}$ $4\overline{)12}$ $4\overline{)52}$ $4\overline{)28}$

S7.18

Multiplication and Division © 2005 • www.zephyrpress.com

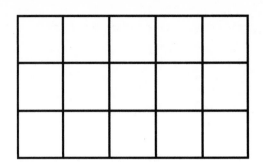

5	2	6	11	8
×5	×8	×4	×8	×12

9	9	15	5	11
×8	×12	×8	×12	×4

10	13	13	15	7
×8	×4	×8	×12	×12

6	8	14	12	4
×12	×8	×4	×4	×4

S7.19

$$\begin{array}{r} 4 \\ \times 12 \\ \hline \end{array} \qquad \begin{array}{r} 9 \\ \times 4 \\ \hline \end{array} \qquad \begin{array}{r} 12 \\ \times 8 \\ \hline \end{array} \qquad \begin{array}{r} 12 \\ \times 12 \\ \hline \end{array} \qquad \begin{array}{r} 3 \\ \times 8 \\ \hline \end{array}$$

$$\begin{array}{r} 4 \\ \times 8 \\ \hline \end{array} \qquad \begin{array}{r} 13 \\ \times 12 \\ \hline \end{array} \qquad \begin{array}{r} 2 \\ \times 4 \\ \hline \end{array} \qquad \begin{array}{r} 14 \\ \times 12 \\ \hline \end{array} \qquad \begin{array}{r} 8 \\ \times 4 \\ \hline \end{array}$$

$$\begin{array}{r} 10 \\ \times 4 \\ \hline \end{array} \qquad \begin{array}{r} 5 \\ \times 8 \\ \hline \end{array} \qquad \begin{array}{r} 11 \\ \times 12 \\ \hline \end{array} \qquad \begin{array}{r} 15 \\ \times 4 \\ \hline \end{array} \qquad \begin{array}{r} 14 \\ \times 8 \\ \hline \end{array}$$

$$\begin{array}{r} 6 \\ \times 8 \\ \hline \end{array} \qquad \begin{array}{r} 7 \\ \times 4 \\ \hline \end{array} \qquad \begin{array}{r} 7 \\ \times 8 \\ \hline \end{array} \qquad \begin{array}{r} 10 \\ \times 12 \\ \hline \end{array} \qquad \begin{array}{r} 3 \\ \times 4 \\ \hline \end{array}$$

S7.19

$$8\overline{)16} \qquad 10\overline{)90} \qquad 12\overline{)84} \qquad 10\overline{)50} \qquad 12\overline{)60}$$

$$4\overline{)52} \qquad 4\overline{)24} \qquad 4\overline{)28} \qquad 4\overline{)36} \qquad 4\overline{)40}$$

$$10\overline{)130} \qquad 12\overline{)108} \qquad 12\overline{)144} \qquad 2\overline{)30} \qquad 12\overline{)132}$$

$$4\overline{)60} \qquad 4\overline{)24} \qquad 4\overline{)12} \qquad 4\overline{)52} \qquad 4\overline{)28}$$

$$8\overline{)56} \qquad 8\overline{)96} \qquad 2\overline{)24} \qquad 12\overline{)72} \qquad 10\overline{)40}$$

S7.19

Name: _____

$4\overline{)32}$　　$4\overline{)8}$　　$4\overline{)20}$　　$4\overline{)16}$　　$4\overline{)12}$

$8\overline{)8}$　　$10\overline{)140}$　　$8\overline{)24}$　　$12\overline{)96}$　　$8\overline{)32}$

$4\overline{)44}$　　$4\overline{)56}$　　$4\overline{)16}$　　$4\overline{)48}$　　$4\overline{)20}$

$8\overline{)40}$　　$12\overline{)24}$　　$2\overline{)28}$　　$12\overline{)36}$　　$2\overline{)18}$

$8\overline{)88}$　　$12\overline{)108}$　　$8\overline{)112}$　　$8\overline{)56}$　　$12\overline{)132}$

S7.19

Multiplication and Division © 2005 • www.zephyrpress.com

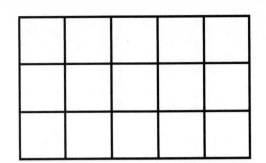

$$\begin{array}{r} 5 \\ \times 6 \\ \hline \end{array}$$
$$\begin{array}{r} 8 \\ \times 6 \\ \hline \end{array}$$
$$\begin{array}{r} 7 \\ \times 6 \\ \hline \end{array}$$
$$\begin{array}{r} 11 \\ \times 6 \\ \hline \end{array}$$
$$\begin{array}{r} 6 \\ \times 6 \\ \hline \end{array}$$

$$\begin{array}{r} 9 \\ \times 6 \\ \hline \end{array}$$
$$\begin{array}{r} 12 \\ \times 6 \\ \hline \end{array}$$
$$\begin{array}{r} 10 \\ \times 6 \\ \hline \end{array}$$
$$\begin{array}{r} 3 \\ \times 6 \\ \hline \end{array}$$
$$\begin{array}{r} 9 \\ \times 6 \\ \hline \end{array}$$

$$\begin{array}{r} 2 \\ \times 6 \\ \hline \end{array}$$
$$\begin{array}{r} 1 \\ \times 6 \\ \hline \end{array}$$
$$\begin{array}{r} 7 \\ \times 6 \\ \hline \end{array}$$
$$\begin{array}{r} 13 \\ \times 6 \\ \hline \end{array}$$
$$\begin{array}{r} 5 \\ \times 6 \\ \hline \end{array}$$

$$\begin{array}{r} 8 \\ \times 6 \\ \hline \end{array}$$
$$\begin{array}{r} 11 \\ \times 6 \\ \hline \end{array}$$
$$\begin{array}{r} 4 \\ \times 6 \\ \hline \end{array}$$
$$\begin{array}{r} 14 \\ \times 6 \\ \hline \end{array}$$
$$\begin{array}{r} 15 \\ \times 6 \\ \hline \end{array}$$

S7.20

Sixes

Division

Alison is having a birthday party. She is giving out party favors to each of her six friends that are coming to her party. Choose a TN of favors that Alison has to give away. Put this number on the large bag. Divide favors evenly into the small bags.

How many favors are in each bag?

Procedure: TN ÷ number of guests

Target numbers: answers on houses chart

From Real to Symbolic:
Use real stickers or wrapped candy for the children to create and solve the problems.

Next, use plastic chips or some other tangible objects to represent the treats.

Last, use number symbols.

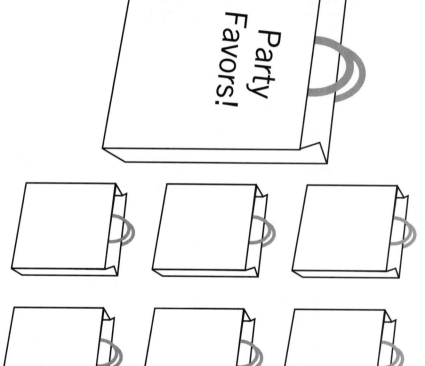

Party Favors!

$$6\overline{)42}$$

7 number in each bag

42 total number of favors

Children

Multiplication

Choose a number of favors for Alison to give to each guest.

How many favors will she need to buy altogether?

Procedure: number of kids X TN (Start with two smaller numbers and work them to arrive at large number.)

Target numbers to choose from:
1–15

S7.21

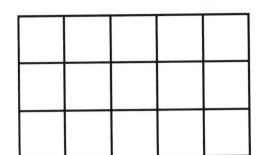

6) 12   6) 78   6) 42   6) 84   6) 36

6) 30   6) 24   6) 60   6) 18   6) 48

6) 54   6) 42   6) 72   6) 90   6) 78

6) 66   6) 36   6) 84   6) 72   6) 54

$$\begin{array}{r} 5 \\ \times 4 \\ \hline \end{array} \qquad \begin{array}{r} 5 \\ \times 6 \\ \hline \end{array} \qquad \begin{array}{r} 3 \\ \times 6 \\ \hline \end{array} \qquad \begin{array}{r} 11 \\ \times 8 \\ \hline \end{array} \qquad \begin{array}{r} 12 \\ \times 6 \\ \hline \end{array}$$

$$\begin{array}{r} 9 \\ \times 8 \\ \hline \end{array} \qquad \begin{array}{r} 13 \\ \times 8 \\ \hline \end{array} \qquad \begin{array}{r} 10 \\ \times 6 \\ \hline \end{array} \qquad \begin{array}{r} 15 \\ \times 8 \\ \hline \end{array} \qquad \begin{array}{r} 4 \\ \times 8 \\ \hline \end{array}$$

$$\begin{array}{r} 10 \\ \times 8 \\ \hline \end{array} \qquad \begin{array}{r} 10 \\ \times 4 \\ \hline \end{array} \qquad \begin{array}{r} 4 \\ \times 6 \\ \hline \end{array} \qquad \begin{array}{r} 11 \\ \times 4 \\ \hline \end{array} \qquad \begin{array}{r} 9 \\ \times 4 \\ \hline \end{array}$$

$$\begin{array}{r} 14 \\ \times 6 \\ \hline \end{array} \qquad \begin{array}{r} 8 \\ \times 8 \\ \hline \end{array} \qquad \begin{array}{r} 14 \\ \times 4 \\ \hline \end{array} \qquad \begin{array}{r} 12 \\ \times 4 \\ \hline \end{array} \qquad \begin{array}{r} 14 \\ \times 8 \\ \hline \end{array}$$

S7.23

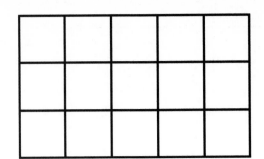

$$\begin{array}{r} 12 \\ \times 8 \\ \hline \end{array} \qquad \begin{array}{r} 11 \\ \times 6 \\ \hline \end{array} \qquad \begin{array}{r} 7 \\ \times 4 \\ \hline \end{array} \qquad \begin{array}{r} 7 \\ \times 8 \\ \hline \end{array} \qquad \begin{array}{r} 9 \\ \times 6 \\ \hline \end{array}$$

$$\begin{array}{r} 2 \\ \times 4 \\ \hline \end{array} \qquad \begin{array}{r} 5 \\ \times 8 \\ \hline \end{array} \qquad \begin{array}{r} 2 \\ \times 6 \\ \hline \end{array} \qquad \begin{array}{r} 13 \\ \times 4 \\ \hline \end{array} \qquad \begin{array}{r} 15 \\ \times 4 \\ \hline \end{array}$$

$$\begin{array}{r} 6 \\ \times 6 \\ \hline \end{array} \qquad \begin{array}{r} 6 \\ \times 4 \\ \hline \end{array} \qquad \begin{array}{r} 15 \\ \times 6 \\ \hline \end{array} \qquad \begin{array}{r} 8 \\ \times 4 \\ \hline \end{array} \qquad \begin{array}{r} 8 \\ \times 6 \\ \hline \end{array}$$

$$\begin{array}{r} 3 \\ \times 4 \\ \hline \end{array} \qquad \begin{array}{r} 13 \\ \times 6 \\ \hline \end{array} \qquad \begin{array}{r} 6 \\ \times 8 \\ \hline \end{array} \qquad \begin{array}{r} 7 \\ \times 6 \\ \hline \end{array} \qquad \begin{array}{r} 4 \\ \times 4 \\ \hline \end{array}$$

$6\overline{)12}$ $6\overline{)78}$ $2\overline{)24}$ $6\overline{)84}$ $10\overline{)40}$

$6\overline{)54}$ $8\overline{)96}$ $6\overline{)42}$ $12\overline{)72}$ $6\overline{)36}$

$8\overline{)56}$ $6\overline{)42}$ $10\overline{)90}$ $10\overline{)50}$ $6\overline{)78}$

$4\overline{)60}$ $6\overline{)72}$ $4\overline{)12}$ $6\overline{)90}$ $12\overline{)60}$

$8\overline{)16}$ $4\overline{)24}$ $12\overline{)84}$ $4\overline{)52}$ $4\overline{)28}$

$6\overline{)30}$ $\qquad$ $4\overline{)36}$ $\qquad$ $6\overline{)60}$ $\qquad$ $6\overline{)18}$ $\qquad$ $2\overline{)16}$

$4\overline{)52}$ $\qquad$ $6\overline{)48}$ $\qquad$ $10\overline{)50}$ $\qquad$ $6\overline{)24}$ $\qquad$ $4\overline{)40}$

$4\overline{)24}$ $\qquad$ $10\overline{)90}$ $\qquad$ $12\overline{)84}$ $\qquad$ $4\overline{)28}$ $\qquad$ $12\overline{)60}$

$6\overline{)36}$ $\qquad$ $12\overline{)24}$ $\qquad$ $2\overline{)28}$ $\qquad$ $6\overline{)72}$ $\qquad$ $2\overline{)18}$

$6\overline{)66}$ $\qquad$ $2\overline{)14}$ $\qquad$ $6\overline{)84}$ $\qquad$ $12\overline{)36}$ $\qquad$ $6\overline{)54}$

Name: _____

$$\begin{array}{r} 5 \\ \times 11 \\ \hline \end{array}\qquad \begin{array}{r} 8 \\ \times 11 \\ \hline \end{array}\qquad \begin{array}{r} 7 \\ \times 11 \\ \hline \end{array}\qquad \begin{array}{r} 11 \\ \times 11 \\ \hline \end{array}\qquad \begin{array}{r} 9 \\ \times 11 \\ \hline \end{array}$$

$$\begin{array}{r} 1 \\ \times 11 \\ \hline \end{array}\qquad \begin{array}{r} 12 \\ \times 11 \\ \hline \end{array}\qquad \begin{array}{r} 10 \\ \times 11 \\ \hline \end{array}\qquad \begin{array}{r} 3 \\ \times 11 \\ \hline \end{array}\qquad \begin{array}{r} 6 \\ \times 11 \\ \hline \end{array}$$

$$\begin{array}{r} 2 \\ \times 11 \\ \hline \end{array}\qquad \begin{array}{r} 9 \\ \times 11 \\ \hline \end{array}\qquad \begin{array}{r} 14 \\ \times 11 \\ \hline \end{array}\qquad \begin{array}{r} 13 \\ \times 11 \\ \hline \end{array}\qquad \begin{array}{r} 5 \\ \times 11 \\ \hline \end{array}$$

$$\begin{array}{r} 8 \\ \times 11 \\ \hline \end{array}\qquad \begin{array}{r} 11 \\ \times 11 \\ \hline \end{array}\qquad \begin{array}{r} 4 \\ \times 11 \\ \hline \end{array}\qquad \begin{array}{r} 7 \\ \times 11 \\ \hline \end{array}\qquad \begin{array}{r} 15 \\ \times 11 \\ \hline \end{array}$$

Multiplication and Division © 2005 • www.zephyrpress.com

Elevens

Arbor Day in Our Neighborhood

Division

We live in a new neighborhood that has 11 houses in it. Because the houses are new, there are no trees yet. All the families went to the nursery together and bought a truckload of trees. (TN)

How many trees will we be able to plant at each house?

Target numbers: answers on houses chart

Procedure: TN ÷ number of homes

From Real to Symbolic:

Cut pictures of trees from magazines or use actual twigs for the children to create and solve the problems.

Next, use plastic chips or some other tangible objects to represent the trees.

Last, use number symbols.

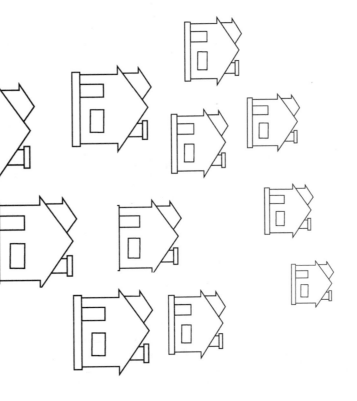

$$\begin{array}{r} 6 \text{ number of trees per home} \\ 11\overline{\smash{\big)}66} \\ \end{array}$$

homes   total number of trees purchased

Multiplication

Choose a number of trees you would like to see planted at each home.

How many trees would you have to buy and haul home in order for every home to have that number of trees?

Procedure: number of homes X TN (Start with two smaller numbers and work them to arrive at large number.)

Target numbers to choose from: 1–15

Multiplication and Division © 2005 • www.zephyrpress.com

11$\overline{)99}$ 11$\overline{)165}$ 11$\overline{)132}$ 11$\overline{)55}$ 11$\overline{)121}$

11$\overline{)154}$ 11$\overline{)33}$ 11$\overline{)110}$ 11$\overline{)77}$ 11$\overline{)143}$

11$\overline{)165}$ 11$\overline{)44}$ 11$\overline{)66}$ 11$\overline{)88}$ 11$\overline{)132}$

11$\overline{)121}$ 11$\overline{)154}$ 11$\overline{)143}$ 11$\overline{)99}$ 11$\overline{)110}$

S7.26

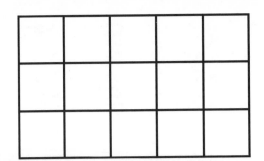

Name: _____

$$\begin{array}{r} 5 \\ \times 5 \\ \hline \end{array}$$
$$\begin{array}{r} 8 \\ \times 5 \\ \hline \end{array}$$
$$\begin{array}{r} 7 \\ \times 5 \\ \hline \end{array}$$
$$\begin{array}{r} 11 \\ \times 5 \\ \hline \end{array}$$
$$\begin{array}{r} 6 \\ \times 5 \\ \hline \end{array}$$

$$\begin{array}{r} 9 \\ \times 5 \\ \hline \end{array}$$
$$\begin{array}{r} 12 \\ \times 5 \\ \hline \end{array}$$
$$\begin{array}{r} 10 \\ \times 5 \\ \hline \end{array}$$
$$\begin{array}{r} 3 \\ \times 5 \\ \hline \end{array}$$
$$\begin{array}{r} 9 \\ \times 5 \\ \hline \end{array}$$

$$\begin{array}{r} 2 \\ \times 5 \\ \hline \end{array}$$
$$\begin{array}{r} 1 \\ \times 5 \\ \hline \end{array}$$
$$\begin{array}{r} 7 \\ \times 5 \\ \hline \end{array}$$
$$\begin{array}{r} 13 \\ \times 5 \\ \hline \end{array}$$
$$\begin{array}{r} 5 \\ \times 5 \\ \hline \end{array}$$

$$\begin{array}{r} 8 \\ \times 5 \\ \hline \end{array}$$
$$\begin{array}{r} 11 \\ \times 5 \\ \hline \end{array}$$
$$\begin{array}{r} 4 \\ \times 5 \\ \hline \end{array}$$
$$\begin{array}{r} 14 \\ \times 5 \\ \hline \end{array}$$
$$\begin{array}{r} 15 \\ \times 5 \\ \hline \end{array}$$

S7.27

Fives

Our Books

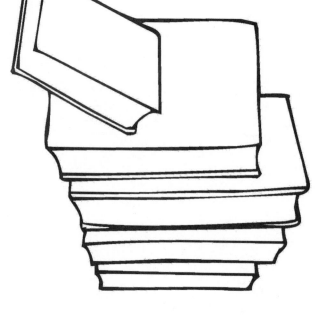

Division

There are five tables in our classroom. Each table has a shelf on the class bookcase. We keep our table's books on our shelf. Choose a TN of books that are on the class bookcase.

How many books are on each shelf?

Target numbers: answers on houses chart

Procedure: TN ÷ number of shelves

From Real to Symbolic:
Use real books for the children to create and solve the problems.

Next, use plastic chips or some other tangible objects to represent the books.

Last, use number symbols.

$$5\overline{)30}$$

6 number on each shelf

30 total number of books

5 shelves

Multiplication

Choose a number of books you would like to see each table make today.

After book-making time is over, all the books will go on the classroom bookcase.

How many books are there altogether?

Procedure: number of shelves X TN (Start with two smaller numbers and work them to arrive at large number.)

Target numbers to choose from: 1–15

S7.28

$5\overline{)60}$ $5\overline{)75}$ $5\overline{)50}$ $5\overline{)35}$ $5\overline{)70}$

$5\overline{)10}$ $5\overline{)40}$ $5\overline{)45}$ $5\overline{)20}$ $5\overline{)60}$

$5\overline{)65}$ $5\overline{)25}$ $5\overline{)20}$ $5\overline{)15}$ $5\overline{)30}$

$5\overline{)35}$ $5\overline{)75}$ $5\overline{)70}$ $5\overline{)40}$ $5\overline{)65}$

S7.29

$$\begin{array}{r} 9 \\ \times 5 \\ \hline \end{array} \qquad \begin{array}{r} 10 \\ \times 11 \\ \hline \end{array} \qquad \begin{array}{r} 14 \\ \times 11 \\ \hline \end{array} \qquad \begin{array}{r} 12 \\ \times 5 \\ \hline \end{array} \qquad \begin{array}{r} 8 \\ \times 11 \\ \hline \end{array} \qquad \begin{array}{r} 11 \\ \times 5 \\ \hline \end{array}$$

$$\begin{array}{r} 10 \\ \times 5 \\ \hline \end{array} \qquad \begin{array}{r} 12 \\ \times 11 \\ \hline \end{array} \qquad \begin{array}{r} 14 \\ \times 5 \\ \hline \end{array} \qquad \begin{array}{r} 11 \\ \times 11 \\ \hline \end{array} \qquad \begin{array}{r} 8 \\ \times 5 \\ \hline \end{array} \qquad \begin{array}{r} 6 \\ \times 11 \\ \hline \end{array}$$

$$\begin{array}{r} 11 \\ \times 5 \\ \hline \end{array} \qquad \begin{array}{r} 13 \\ \times 5 \\ \hline \end{array} \qquad \begin{array}{r} 13 \\ \times 11 \\ \hline \end{array} \qquad \begin{array}{r} 3 \\ \times 5 \\ \hline \end{array} \qquad \begin{array}{r} 15 \\ \times 11 \\ \hline \end{array} \qquad \begin{array}{r} 7 \\ \times 11 \\ \hline \end{array}$$

$$\begin{array}{r} 7 \\ \times 5 \\ \hline \end{array} \qquad \begin{array}{r} 3 \\ \times 11 \\ \hline \end{array} \qquad \begin{array}{r} 15 \\ \times 5 \\ \hline \end{array} \qquad \begin{array}{r} 6 \\ \times 5 \\ \hline \end{array} \qquad \begin{array}{r} 9 \\ \times 11 \\ \hline \end{array} \qquad \begin{array}{r} 5 \\ \times 5 \\ \hline \end{array}$$

S7.30

$5\overline{)30}$ $11\overline{)132}$ $11\overline{)121}$ $5\overline{)45}$ $5\overline{)60}$

$11\overline{)33}$ $5\overline{)40}$ $5\overline{)50}$ $5\overline{)75}$ $11\overline{)110}$

$5\overline{)20}$ $11\overline{)154}$ $5\overline{)35}$ $11\overline{)99}$ $11\overline{)55}$

$5\overline{)65}$ $5\overline{)25}$ $5\overline{)20}$ $5\overline{)15}$ $5\overline{)70}$

S7.30

$$\begin{array}{r} 5 \\ \times 9 \\ \hline \end{array} \qquad \begin{array}{r} 8 \\ \times 9 \\ \hline \end{array} \qquad \begin{array}{r} 7 \\ \times 9 \\ \hline \end{array} \qquad \begin{array}{r} 11 \\ \times 9 \\ \hline \end{array} \qquad \begin{array}{r} 6 \\ \times 9 \\ \hline \end{array}$$

$$\begin{array}{r} 9 \\ \times 9 \\ \hline \end{array} \qquad \begin{array}{r} 12 \\ \times 9 \\ \hline \end{array} \qquad \begin{array}{r} 10 \\ \times 9 \\ \hline \end{array} \qquad \begin{array}{r} 3 \\ \times 9 \\ \hline \end{array} \qquad \begin{array}{r} 9 \\ \times 9 \\ \hline \end{array}$$

$$\begin{array}{r} 2 \\ \times 9 \\ \hline \end{array} \qquad \begin{array}{r} 1 \\ \times 9 \\ \hline \end{array} \qquad \begin{array}{r} 7 \\ \times 9 \\ \hline \end{array} \qquad \begin{array}{r} 13 \\ \times 9 \\ \hline \end{array} \qquad \begin{array}{r} 5 \\ \times 9 \\ \hline \end{array}$$

$$\begin{array}{r} 8 \\ \times 9 \\ \hline \end{array} \qquad \begin{array}{r} 11 \\ \times 9 \\ \hline \end{array} \qquad \begin{array}{r} 4 \\ \times 9 \\ \hline \end{array} \qquad \begin{array}{r} 14 \\ \times 9 \\ \hline \end{array} \qquad \begin{array}{r} 15 \\ \times 9 \\ \hline \end{array}$$

S7.31

Nines

Our Music Store

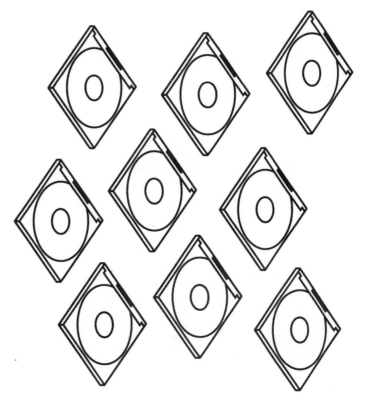

Division

My friends and I are collecting used CDs to sell in our store. We made 9 bins out of milk crates to hold all the CDs we are going to find. We have a pile of (TN) CDs to put in our bins.

How many CDs will be put in each of our 9 bins?

Target numbers: answers on houses chart

Procedure: TN ÷ number of bins

From Real to Symbolic:

Use pieces of square cardboard or even old CD cases if you can find some for the children to create and solve the problems.

Next, use plastic chips or some other tangible objects to represent the CDs

Last, use number symbols.

$$9\,\overline{)81}\ \ \substack{\text{9 number of CDs in each bin}\\ \text{total number of CDs}}$$

bins

Multiplication

Choose a number of CDs you think would fit nicely in each CD bin.

How many CDs would you have to find in order to put that many in each bin?

Procedure: number of bins X TN (Start with two smaller numbers and work them to arrive at large number.)

Target numbers to choose from: 1–15

9 | 18   9 | 45   9 | 81   9 | 63   9 | 72

9 | 27   9 | 36   9 | 108   9 | 54   9 | 126

9 | 63   9 | 81   9 | 90   9 | 45   9 | 135

9 | 72   9 | 36   9 | 54   9 | 117   9 | 27

S7.33

Multiplication and Division © 2005 • www.zephyrpress.com

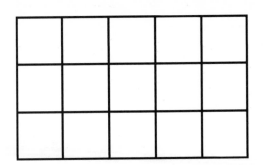

9	14	12	12	11	7
×9	×11	×5	×9	×11	×5

6	12	13	7	8	4
×9	×11	×9	×9	×5	×9

15	3	13	13	14	15
×9	×9	×5	×11	×9	×11

8	5	15	6	14	5
×9	×9	×5	×5	×5	×5

$9\overline{)18}$ $11\overline{)121}$ $9\overline{)45}$ $5\overline{)30}$ $9\overline{)63}$

$9\overline{)27}$ $9\overline{)36}$ $5\overline{)40}$ $9\overline{)54}$ $5\overline{)45}$

$9\overline{)81}$ $5\overline{)50}$ $11\overline{)132}$ $5\overline{)35}$ $9\overline{)72}$

$9\overline{)108}$ $5\overline{)20}$ $11\overline{)33}$ $5\overline{)60}$ $9\overline{)90}$

S7.34

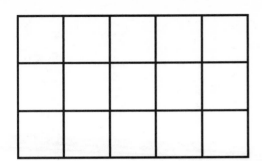

Name: _____

$$\begin{array}{r} 9 \\ \times 3 \\ \hline \end{array} \qquad \begin{array}{r} 8 \\ \times 3 \\ \hline \end{array} \qquad \begin{array}{r} 5 \\ \times 3 \\ \hline \end{array} \qquad \begin{array}{r} 11 \\ \times 3 \\ \hline \end{array} \qquad \begin{array}{r} 2 \\ \times 3 \\ \hline \end{array}$$

$$\begin{array}{r} 7 \\ \times 3 \\ \hline \end{array} \qquad \begin{array}{r} 12 \\ \times 3 \\ \hline \end{array} \qquad \begin{array}{r} 10 \\ \times 3 \\ \hline \end{array} \qquad \begin{array}{r} 3 \\ \times 3 \\ \hline \end{array} \qquad \begin{array}{r} 4 \\ \times 3 \\ \hline \end{array}$$

$$\begin{array}{r} 13 \\ \times 3 \\ \hline \end{array} \qquad \begin{array}{r} 2 \\ \times 3 \\ \hline \end{array} \qquad \begin{array}{r} 9 \\ \times 3 \\ \hline \end{array} \qquad \begin{array}{r} 5 \\ \times 3 \\ \hline \end{array} \qquad \begin{array}{r} 14 \\ \times 3 \\ \hline \end{array}$$

$$\begin{array}{r} 8 \\ \times 3 \\ \hline \end{array} \qquad \begin{array}{r} 11 \\ \times 3 \\ \hline \end{array} \qquad \begin{array}{r} 4 \\ \times 3 \\ \hline \end{array} \qquad \begin{array}{r} 15 \\ \times 3 \\ \hline \end{array} \qquad \begin{array}{r} 7 \\ \times 3 \\ \hline \end{array}$$

S7.35

Threes

Division

Ed, Ted, and Fred are triplets who love to play marbles together. They keep their marbles altogether in a big bag. When they play, they each get the same number of marbles.

Choose a target number (TN) of marbles they will play with and put that number on the large bag.

Give each child the same number of marbles. How many do they each get?

Procedure: TN ÷ number of kids.

Target numbers: answers on houses chart

From Real to Symbolic:
Use real marbles for the children to create and solve the problems.

Next, use plastic chips or some other tangible objects to represent the marbles.

Last, use number symbols.

Our Marbles

$$\frac{11 \text{ number for each child}}{3 \overline{\smash{)}33}}$$

kids — total number of marbles

Ed

Ted

Fred

Multiplication

When the boys play marbles, they each get a TN of marbles to play with. Choose the number of marbles you want each boy to have. Let them play. When they finish, they will put them altogether.

How many marbles are there altogether?

Procedure: number of kids X TN. (Start with two smaller numbers and work them to arrive at large number.)

Target numbers to choose from:
1–15

S7.36

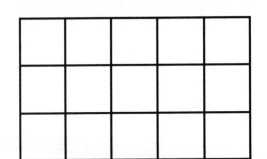

$3\overline{)18}$ $3\overline{)6}$ $3\overline{)21}$ $3\overline{)15}$ $3\overline{)12}$

$3\overline{)30}$ $3\overline{)24}$ $3\overline{)27}$ $3\overline{)33}$ $3\overline{)39}$

$3\overline{)9}$ $3\overline{)36}$ $3\overline{)42}$ $3\overline{)45}$ $3\overline{)18}$

$3\overline{)27}$ $3\overline{)15}$ $3\overline{)21}$ $3\overline{)24}$ $3\overline{)12}$

S7.37

$$\begin{array}{r} 7 \\ \times 3 \\ \hline \end{array}$$
$$\begin{array}{r} 14 \\ \times 3 \\ \hline \end{array}$$
$$\begin{array}{r} 14 \\ \times 11 \\ \hline \end{array}$$
$$\begin{array}{r} 12 \\ \times 5 \\ \hline \end{array}$$
$$\begin{array}{r} 12 \\ \times 11 \\ \hline \end{array}$$
$$\begin{array}{r} 11 \\ \times 11 \\ \hline \end{array}$$

$$\begin{array}{r} 13 \\ \times 9 \\ \hline \end{array}$$
$$\begin{array}{r} 8 \\ \times 11 \\ \hline \end{array}$$
$$\begin{array}{r} 7 \\ \times 9 \\ \hline \end{array}$$
$$\begin{array}{r} 12 \\ \times 9 \\ \hline \end{array}$$
$$\begin{array}{r} 9 \\ \times 3 \\ \hline \end{array}$$
$$\begin{array}{r} 12 \\ \times 3 \\ \hline \end{array}$$

$$\begin{array}{r} 15 \\ \times 9 \\ \hline \end{array}$$
$$\begin{array}{r} 13 \\ \times 3 \\ \hline \end{array}$$
$$\begin{array}{r} 8 \\ \times 3 \\ \hline \end{array}$$
$$\begin{array}{r} 13 \\ \times 11 \\ \hline \end{array}$$
$$\begin{array}{r} 15 \\ \times 11 \\ \hline \end{array}$$
$$\begin{array}{r} 6 \\ \times 3 \\ \hline \end{array}$$

$$\begin{array}{r} 8 \\ \times 9 \\ \hline \end{array}$$
$$\begin{array}{r} 5 \\ \times 3 \\ \hline \end{array}$$
$$\begin{array}{r} 12 \\ \times 3 \\ \hline \end{array}$$
$$\begin{array}{r} 4 \\ \times 3 \\ \hline \end{array}$$
$$\begin{array}{r} 15 \\ \times 5 \\ \hline \end{array}$$
$$\begin{array}{r} 14 \\ \times 9 \\ \hline \end{array}$$

S7.38

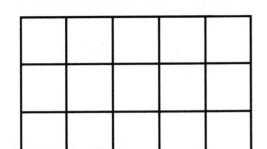

Name: _____

$3\overline{)18}$ $9\overline{)27}$ $3\overline{)21}$ $3\overline{)15}$ $3\overline{)12}$

$3\overline{)30}$ $3\overline{)24}$ $3\overline{)27}$ $9\overline{)36}$ $3\overline{)39}$

$3\overline{)9}$ $11\overline{)33}$ $3\overline{)42}$ $9\overline{)54}$ $3\overline{)33}$

$9\overline{)108}$ $3\overline{)36}$ $9\overline{)81}$ $3\overline{)45}$ $3\overline{)18}$

$$\begin{array}{r} 5 \\ \times 7 \\ \hline \end{array} \quad \begin{array}{r} 8 \\ \times 7 \\ \hline \end{array} \quad \begin{array}{r} 7 \\ \times 7 \\ \hline \end{array} \quad \begin{array}{r} 11 \\ \times 7 \\ \hline \end{array} \quad \begin{array}{r} 6 \\ \times 7 \\ \hline \end{array}$$

$$\begin{array}{r} 9 \\ \times 7 \\ \hline \end{array} \quad \begin{array}{r} 12 \\ \times 7 \\ \hline \end{array} \quad \begin{array}{r} 10 \\ \times 7 \\ \hline \end{array} \quad \begin{array}{r} 3 \\ \times 7 \\ \hline \end{array} \quad \begin{array}{r} 9 \\ \times 7 \\ \hline \end{array}$$

$$\begin{array}{r} 2 \\ \times 7 \\ \hline \end{array} \quad \begin{array}{r} 1 \\ \times 7 \\ \hline \end{array} \quad \begin{array}{r} 7 \\ \times 7 \\ \hline \end{array} \quad \begin{array}{r} 13 \\ \times 7 \\ \hline \end{array} \quad \begin{array}{r} 5 \\ \times 7 \\ \hline \end{array}$$

$$\begin{array}{r} 8 \\ \times 7 \\ \hline \end{array} \quad \begin{array}{r} 11 \\ \times 7 \\ \hline \end{array} \quad \begin{array}{r} 4 \\ \times 7 \\ \hline \end{array} \quad \begin{array}{r} 14 \\ \times 7 \\ \hline \end{array} \quad \begin{array}{r} 15 \\ \times 7 \\ \hline \end{array}$$

Sevens

Multiplication

Choose a number of plants you would like to plant in each row. How many plants will you need altogether?

Procedure: number of rows X TN (Start with two smaller numbers and work them to arrive at large number.)

Target numbers to choose from: 1–15

Division

We are planting a garden. We've sprouted TN of seedlings that we need to plant outside now. We have our ground all ready, and we have made seven rows to plant in. We want to plant the same number of plants in each row.

How many plants are in each row?

Procedure: TN ÷ number of rows

Target numbers: answers on houses chart

From Real to Symbolic:
Use leaves or pictures of plants for the children to create and solve the problems.

Next, use plastic chips or some other tangible objects to represent the plants.

Last, use number symbols.

Our Garden

$$\begin{array}{r} 10 \text{ number in each row} \\ 7\overline{)70} \text{ total number of plants} \\ \text{rows} \end{array}$$

S7.40

$7\overline{)14}$ $7\overline{)49}$ $7\overline{)63}$ $7\overline{)35}$ $7\overline{)84}$

$7\overline{)42}$ $7\overline{)56}$ $7\overline{)28}$ $7\overline{)105}$ $7\overline{)98}$

$7\overline{)21}$ $7\overline{)91}$ $7\overline{)77}$ $7\overline{)63}$ $7\overline{)35}$

$7\overline{)70}$ $7\overline{)49}$ $7\overline{)56}$ $7\overline{)42}$ $7\overline{)84}$

S7.41

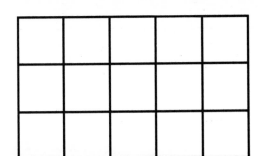

Name: _____

$$\begin{array}{r} 8 \\ \times 3 \\ \hline \end{array}$$
$$\begin{array}{r} 14 \\ \times 3 \\ \hline \end{array}$$
$$\begin{array}{r} 7 \\ \times 7 \\ \hline \end{array}$$
$$\begin{array}{r} 12 \\ \times 5 \\ \hline \end{array}$$
$$\begin{array}{r} 9 \\ \times 9 \\ \hline \end{array}$$
$$\begin{array}{r} 12 \\ \times 9 \\ \hline \end{array}$$

$$\begin{array}{r} 7 \\ \times 9 \\ \hline \end{array}$$
$$\begin{array}{r} 3 \\ \times 7 \\ \hline \end{array}$$
$$\begin{array}{r} 9 \\ \times 7 \\ \hline \end{array}$$
$$\begin{array}{r} 13 \\ \times 9 \\ \hline \end{array}$$
$$\begin{array}{r} 14 \\ \times 7 \\ \hline \end{array}$$
$$\begin{array}{r} 12 \\ \times 3 \\ \hline \end{array}$$

$$\begin{array}{r} 15 \\ \times 9 \\ \hline \end{array}$$
$$\begin{array}{r} 13 \\ \times 3 \\ \hline \end{array}$$
$$\begin{array}{r} 15 \\ \times 7 \\ \hline \end{array}$$
$$\begin{array}{r} 8 \\ \times 9 \\ \hline \end{array}$$
$$\begin{array}{r} 9 \\ \times 3 \\ \hline \end{array}$$
$$\begin{array}{r} 8 \\ \times 7 \\ \hline \end{array}$$

$$\begin{array}{r} 13 \\ \times 7 \\ \hline \end{array}$$
$$\begin{array}{r} 5 \\ \times 9 \\ \hline \end{array}$$
$$\begin{array}{r} 15 \\ \times 5 \\ \hline \end{array}$$
$$\begin{array}{r} 12 \\ \times 7 \\ \hline \end{array}$$
$$\begin{array}{r} 6 \\ \times 7 \\ \hline \end{array}$$
$$\begin{array}{r} 14 \\ \times 9 \\ \hline \end{array}$$

S7.42

$7\overline{)28}$ $9\overline{)45}$ $7\overline{)49}$ $5\overline{)50}$ $7\overline{)14}$

$5\overline{)30}$ $7\overline{)42}$ $5\overline{)40}$ $7\overline{)63}$ $9\overline{)18}$

$5\overline{)60}$ $7\overline{)56}$ $11\overline{)121}$ $7\overline{)21}$ $7\overline{)91}$

$5\overline{)20}$ $7\overline{)35}$ $5\overline{)35}$ $7\overline{)84}$ $11\overline{)132}$

Name: _____

Solve.

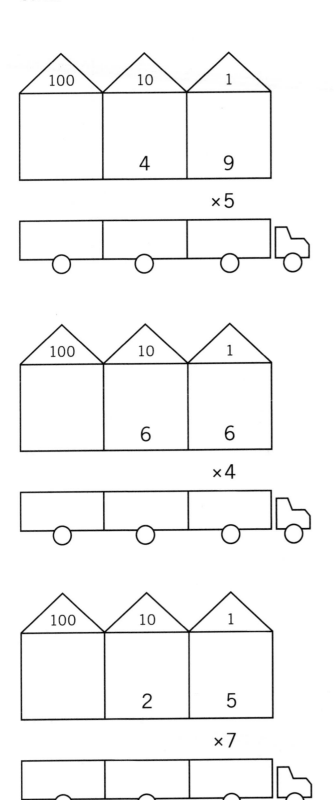

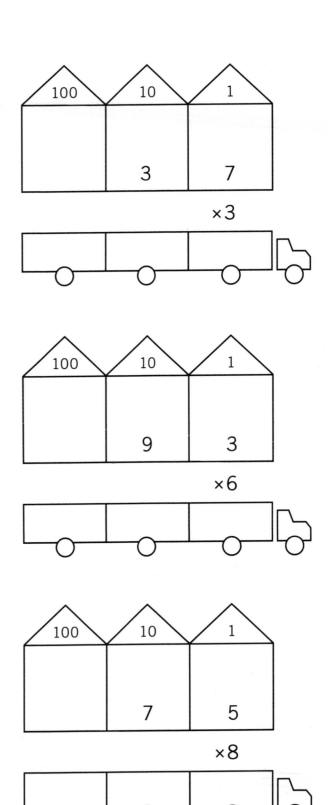

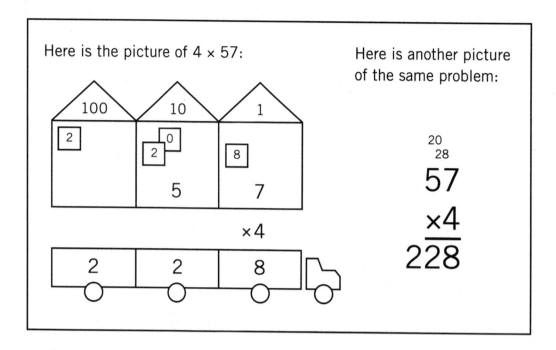

Here is the picture of 4 × 57:

Here is another picture of the same problem:

$$\begin{array}{r} \overset{20}{\underset{}{\overset{28}{}}} \\ 57 \\ \times 4 \\ \hline 228 \end{array}$$

Use the example above as a model for solving these problems:

$$\begin{array}{r} 68 \\ \times 3 \\ \hline \end{array} \qquad \begin{array}{r} 24 \\ \times 7 \\ \hline \end{array} \qquad \begin{array}{r} 47 \\ \times 9 \\ \hline \end{array} \qquad \begin{array}{r} 23 \\ \times 6 \\ \hline \end{array} \qquad \begin{array}{r} 46 \\ \times 5 \\ \hline \end{array} \qquad \begin{array}{r} 57 \\ \times 4 \\ \hline \end{array}$$

$$\begin{array}{r} 43 \\ \times 8 \\ \hline \end{array} \qquad \begin{array}{r} 87 \\ \times 2 \\ \hline \end{array} \qquad \begin{array}{r} 48 \\ \times 3 \\ \hline \end{array} \qquad \begin{array}{r} 88 \\ \times 7 \\ \hline \end{array} \qquad \begin{array}{r} 65 \\ \times 9 \\ \hline \end{array} \qquad \begin{array}{r} 42 \\ \times 6 \\ \hline \end{array}$$

$$\begin{array}{r} 25 \\ \times 5 \\ \hline \end{array} \qquad \begin{array}{r} 37 \\ \times 4 \\ \hline \end{array} \qquad \begin{array}{r} 44 \\ \times 8 \\ \hline \end{array} \qquad \begin{array}{r} 77 \\ \times 2 \\ \hline \end{array} \qquad \begin{array}{r} 56 \\ \times 3 \\ \hline \end{array} \qquad \begin{array}{r} 46 \\ \times 7 \\ \hline \end{array}$$

S8.2

Multiplication and Division © 2005 • www.zephyrpress.com

Solve. Notice that you can load the truck as
you go. The first problem is done for you.

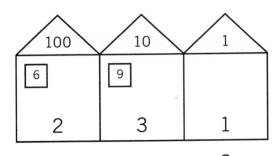

×3

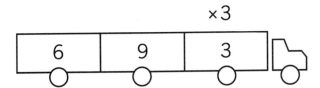

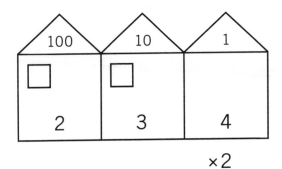

×2

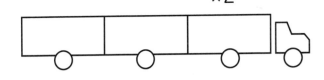

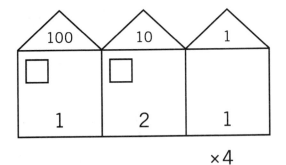

×4

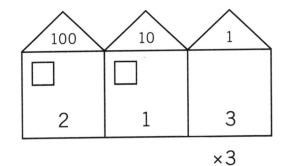

×3

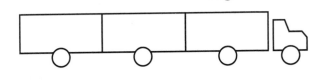

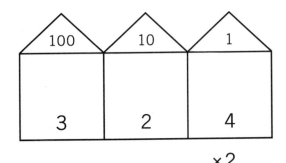

×2

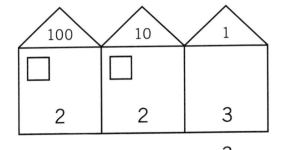

×3

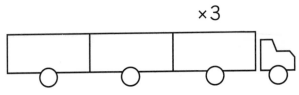

Solve.

758	443	452	257	655
×3	×9	×5	×8	×3

359	495	786	348	654
×6	×4	×2	×7	×4

884	247	556	254	548
×6	×8	×5	×9	×7

437	295	649	753	573
×5	×8	×9	×6	×7

Name: _____

Solve.

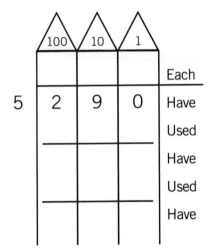

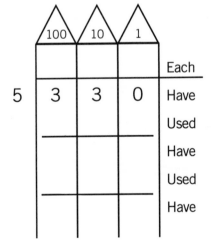

Solve.

$2\overline{)110}$ $5\overline{)225}$ $4\overline{)112}$ $8\overline{)136}$ $3\overline{)252}$

$3\overline{)285}$ $9\overline{)207}$ $6\overline{)192}$ $7\overline{)252}$ $2\overline{)126}$

$8\overline{)528}$ $5\overline{)255}$ $4\overline{)196}$ $9\overline{)387}$ $7\overline{)112}$

$2\overline{)154}$ $4\overline{)228}$ $3\overline{)201}$ $2\overline{)134}$ $6\overline{)312}$

S9.2

Solve.

$2 \overline{)158}$ $5 \overline{)295}$ $4 \overline{)332}$ $8 \overline{)496}$ $3 \overline{)222}$

$3 \overline{)207}$ $9 \overline{)522}$ $6 \overline{)504}$ $7 \overline{)525}$ $2 \overline{)178}$

$8 \overline{)592}$ $5 \overline{)345}$ $4 \overline{)272}$ $9 \overline{)603}$ $7 \overline{)385}$

$2 \overline{)198}$ $4 \overline{)352}$ $3 \overline{)267}$ $2 \overline{)194}$ $6 \overline{)492}$

Name: _____

Solve. You will have a remainder on these problems. The first one is done for you.

$$
\begin{array}{r}
58r1 \\
2\overline{)117} \\
-10 \\
\hline
17 \\
-16 \\
\hline
1
\end{array}
$$
$5\overline{)229}$ $4\overline{)115}$ $8\overline{)139}$ $3\overline{)255}$

$3\overline{)238}$ $9\overline{)208}$ $6\overline{)193}$ $7\overline{)155}$ $2\overline{)128}$

$8\overline{)658}$ $5\overline{)256}$ $4\overline{)197}$ $9\overline{)389}$ $7\overline{)114}$

$2\overline{)154}$ $4\overline{)228}$ $3\overline{)201}$ $2\overline{)134}$ $6\overline{)312}$

Multiplication and Division © 2005 • www.zephyrpress.com

Name: _____

Solve.

$2\overline{)1310}$ $\qquad$ $5\overline{)2025}$ $\qquad$ $4\overline{)1322}$ $\qquad$ $8\overline{)1336}$

$3\overline{)2185}$ $\qquad$ $9\overline{)2307}$ $\qquad$ $6\overline{)1592}$ $\qquad$ $7\overline{)2152}$

$8\overline{)5228}$ $\qquad$ $5\overline{)2655}$ $\qquad$ $4\overline{)1946}$ $\qquad$ $9\overline{)3187}$

S9.5

Solve.

$2 \overline{)1538}$ $5 \overline{)2385}$ $4 \overline{)3235}$ $8 \overline{)2212}$

$3 \overline{)2057}$ $9 \overline{)5223}$ $6 \overline{)5245}$ $7 \overline{)1783}$

$8 \overline{)5834}$ $5 \overline{)4567}$ $4 \overline{)2573}$ $9 \overline{)2493}$

S9.6

Solve.

$2 \overline{)110}$ $5 \overline{)225}$ $4 \overline{)112}$ $8 \overline{)136}$ $3 \overline{)252}$

$4 \overline{)296}$ $9 \overline{)127}$ $8 \overline{)291}$ $7 \overline{)257}$ $2 \overline{)351}$

$6 \overline{)248}$ $5 \overline{)255}$ $3 \overline{)285}$ $9 \overline{)387}$ $7 \overline{)342}$

$2 \overline{)574}$ $4 \overline{)458}$ $6 \overline{)335}$ $2 \overline{)174}$ $6 \overline{)201}$

S9.7

Solve.

$$2\overline{)158} \quad 5\overline{)295} \quad 4\overline{)332} \quad 8\overline{)496} \quad 3\overline{)222}$$

$$3\overline{)253} \quad 9\overline{)598} \quad 6\overline{)243} \quad 7\overline{)527} \quad 2\overline{)158}$$

$$8\overline{)571} \quad 5\overline{)158} \quad 4\overline{)243} \quad 9\overline{)548} \quad 7\overline{)125}$$

$$2\overline{)541} \quad 4\overline{)287} \quad 3\overline{)357} \quad 2\overline{)243} \quad 6\overline{)582}$$

S9.8

Teacher Assessments and Overheads

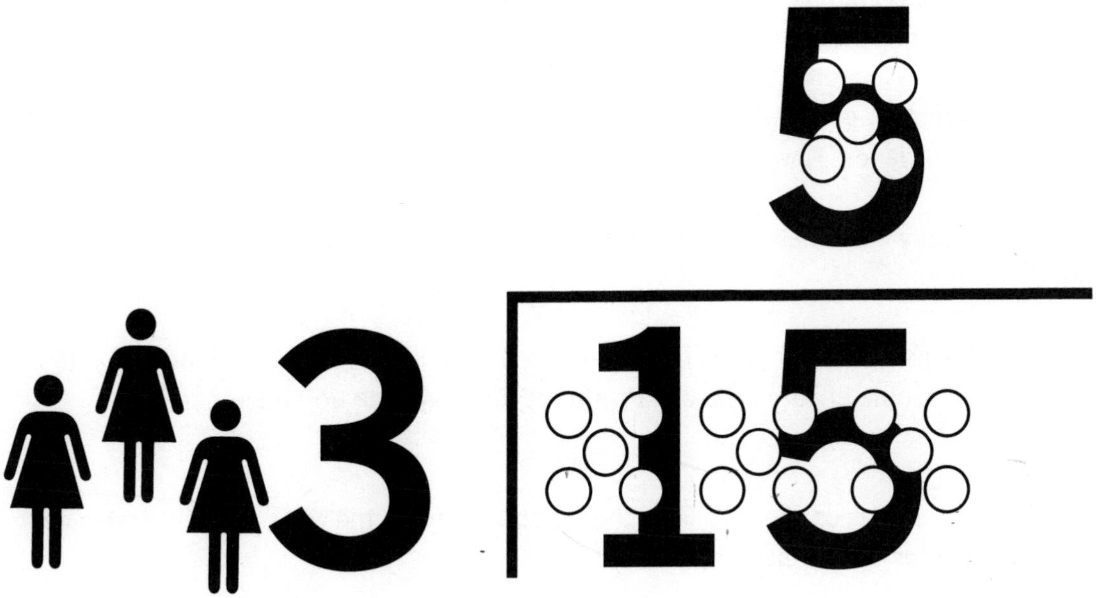

T2.1

Name: _____

A. Read each story and decide if you need to multiply or divide to find the answer. Put an "M" for multiply or a "D" for divide on the line in front of each story.

_____1. Three boys had four yo-yos each. How many yo-yos are there in all?

_____2. Jaden told Darien he would help him deliver his papers. If Darien delivers papers to 30 houses, how many papers would each boy deliver?

_____3. Erica's mother gave Erica a bag of 24 gum balls and told her to share them fairly among herself and her three friends. How many gum balls did each of the four girls get?

_____4. Each family on Sweet Street has three pets. If there are seven houses on the street, how many pets are there in all?

_____5. After Sasha's slumber party, her mom asked each child to pick up four things to put away. If there were nine children, how many things did they pick up altogether?

B. Read the word stories on the left. Find the number story on the right that illustrates each story. Put the correct letter on the blank by each word story.

_____1. Jordan had a collection of 35 matchbox cars. He had four friends over to play. If he shared the cars equally with his friends, how many cars would each of the five boys get?

A. $9 \times 9 = 81$

_____2. Jillian's aunt gave her 24 cat stickers to share with her three sisters. How many stickers did each of the four girls get?

B. $3 \times 7 = 21$

_____3. Nine friends each had nine marbles. They decided to put them together. How many marbles were in the pile they made?

C. $4\overline{)24}$ (with 6 above)

_____4. Kara's three cats each had seven kittens. How many kittens did Kara have in all?

D. $4 \times 9 = 36$

_____5. Tom and his friends were washing cars at a car wash. Tom's job was to wash the tires. Before lunch, Tom had washed the tires on nine cars. How many tires did he wash in all?

C. $5\overline{)35}$ (with 7 above)

Concept Mastery Tracking Chart

Names:	Concepts of M/D	Complete Map	Address Locations	10s Facts M/D	2s Facts M/D	12s Facts M/D	8s Facts M/D	4s Facts M/D	6s Facts M/D	11s Facts M/D	5s Facts M/D	9s Facts M/D	3s Facts M/D	7s Facts M/D	Assess Evens M/D	Assess Odds M/D	Assess Final: M	Assess Final: D	Assess Mixed M/D	Multi-digit M	Assess	Multi-digit D	Assess

1				

2				

3				

4				

5				

6				

7				

8				

9				

10				

11				

12				

Find the Addresses

Read each problem and draw a line to the box that
shows that problem. The first one is done for you.

$3 \times 7 =$

$5 \times 9 =$

$1 \times 10 =$

$9 \times 4 =$

$7 \times 8 =$

$11 \times 3 =$

$4 \times 8 =$

$2 \times 11 =$

$8 \times 9 =$

$6 \times 13 =$

$12 \times 4 =$

$10 \times 14 =$

1	2	3	4	5
6	7	8	9	10
11	12	13	14	15

2	4	6	8	10
12	14	16	18	20
22	24	26	28	30

3	6	9	12	15
18	21	24	27	30
33	36	39	42	45

4	8	12	16	20
24	28	32	36	40
44	48	52	56	60

5	10	15	20	25
30	35	40	45	50
55	60	65	70	75

6	12	18	24	30
36	42	48	54	60
66	72	78	84	90

7	14	21	28	35
42	49	56	63	70
77	84	91	98	105

8	16	24	32	40
48	56	64	72	80
88	96	104	112	120

9	18	27	36	45
54	63	72	81	90
99	108	117	126	135

10	20	30	40	50
60	70	80	90	100
110	120	130	140	150

11	22	33	44	55
66	77	88	99	110
121	132	143	154	165

12	24	36	48	60
72	84	96	108	120
132	144	156	168	180

T4.1

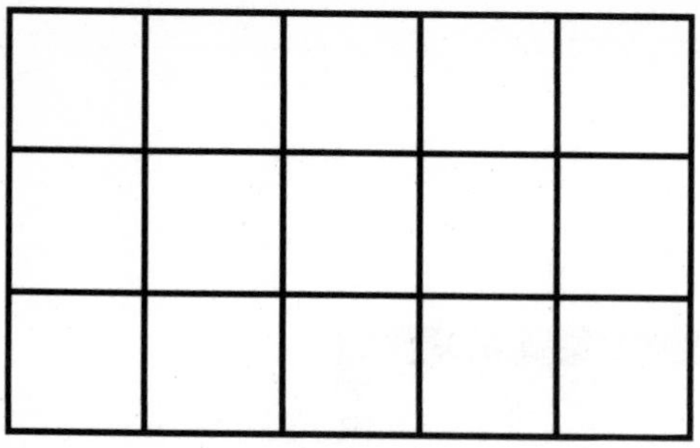

Name: ______________________________

14 ×10	5 ×8	7 ×10	9 ×2	6 ×10	5 ×4
9 ×10	9 ×4	12 ×12	3 ×10	7 ×2	5 ×6
15 ×12	10 ×4	11 ×10	11 ×4	11 ×2	12 ×4
14 ×6	8 ×8	14 ×4	15 ×6	14 ×8	7 ×4

T7.1

$2\overline{)14}$ $6\overline{)36}$ $12\overline{)156}$ $10\overline{)90}$ $4\overline{)52}$

$2\overline{)20}$ $6\overline{)84}$ $10\overline{)110}$ $8\overline{)96}$ $4\overline{)44}$

$8\overline{)72}$ $6\overline{)42}$ $4\overline{)56}$ $10\overline{)100}$ $8\overline{)64}$

$6\overline{)48}$ $12\overline{)24}$ $4\overline{)24}$ $2\overline{)30}$ $12\overline{)144}$

T7.1

$$14 \times 2 \qquad 15 \times 4 \qquad 14 \times 12 \qquad 7 \times 6 \qquad 3 \times 4 \qquad 3 \times 12$$

$$10 \times 12 \qquad 6 \times 12 \qquad 6 \times 2 \qquad 11 \times 12 \qquad 8 \times 5 \qquad 3 \times 6$$

$$5 \times 12 \qquad 7 \times 8 \qquad 3 \times 2 \qquad 10 \times 8 \qquad 8 \times 12 \qquad 7 \times 12$$

$$9 \times 6 \qquad 8 \times 4 \qquad 9 \times 8 \qquad 12 \times 2 \qquad 11 \times 6 \qquad 11 \times 8$$

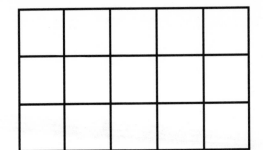

Name: _____

$6\overline{)18}$ $2\overline{)6}$ $10\overline{)70}$ $2\overline{)10}$ $8\overline{)80}$

$4\overline{)28}$ $6\overline{)78}$ $4\overline{)16}$ $10\overline{)60}$ $8\overline{)48}$

$2\overline{)18}$ $10\overline{)140}$ $12\overline{)72}$ $6\overline{)12}$ $4\overline{)40}$

$8\overline{)112}$ $6\overline{)60}$ $12\overline{)108}$ $10\overline{)90}$ $8\overline{)16}$

$$\begin{array}{r} 8 \\ \times 6 \\ \hline \end{array} \qquad \begin{array}{r} 13 \\ \times 8 \\ \hline \end{array} \qquad \begin{array}{r} 6 \\ \times 4 \\ \hline \end{array} \qquad \begin{array}{r} 5 \\ \times 10 \\ \hline \end{array} \qquad \begin{array}{r} 13 \\ \times 12 \\ \hline \end{array} \qquad \begin{array}{r} 3 \\ \times 8 \\ \hline \end{array}$$

$$\begin{array}{r} 13 \\ \times 4 \\ \hline \end{array} \qquad \begin{array}{r} 13 \\ \times 6 \\ \hline \end{array} \qquad \begin{array}{r} 4 \\ \times 8 \\ \hline \end{array} \qquad \begin{array}{r} 12 \\ \times 10 \\ \hline \end{array} \qquad \begin{array}{r} 4 \\ \times 12 \\ \hline \end{array} \qquad \begin{array}{r} 15 \\ \times 10 \\ \hline \end{array}$$

$$\begin{array}{r} 15 \\ \times 8 \\ \hline \end{array} \qquad \begin{array}{r} 10 \\ \times 10 \\ \hline \end{array} \qquad \begin{array}{r} 2 \\ \times 4 \\ \hline \end{array} \qquad \begin{array}{r} 10 \\ \times 6 \\ \hline \end{array} \qquad \begin{array}{r} 15 \\ \times 2 \\ \hline \end{array} \qquad \begin{array}{r} 13 \\ \times 10 \\ \hline \end{array}$$

$$\begin{array}{r} 9 \\ \times 12 \\ \hline \end{array} \qquad \begin{array}{r} 12 \\ \times 6 \\ \hline \end{array} \qquad \begin{array}{r} 6 \\ \times 8 \\ \hline \end{array} \qquad \begin{array}{r} 4 \\ \times 6 \\ \hline \end{array} \qquad \begin{array}{r} 8 \\ \times 10 \\ \hline \end{array} \qquad \begin{array}{r} 6 \\ \times 6 \\ \hline \end{array}$$

T7.3

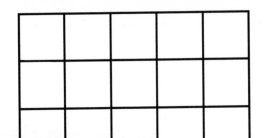

Name: _____

$2\overline{)12}$ $8\overline{)56}$ $12\overline{)48}$ $6\overline{)30}$ $10\overline{)150}$

$4\overline{)36}$ $2\overline{)8}$ $8\overline{)40}$ $12\overline{)120}$ $4\overline{)48}$

$6\overline{)24}$ $12\overline{)96}$ $10\overline{)20}$ $2\overline{)22}$ $10\overline{)130}$

$2\overline{)24}$ $10\overline{)60}$ $8\overline{)40}$ $12\overline{)168}$ $4\overline{)20}$

T7.3

| 8 | 14 | 7 | 4 | 7 | 12 |
| ×3 | ×3 | ×7 | ×3 | ×9 | ×9 |

| 13 | 3 | 9 | 13 | 13 | 15 |
| ×9 | ×7 | ×9 | ×11 | ×7 | ×3 |

| 12 | 14 | 13 | 8 | 9 | 5 |
| ×3 | ×7 | ×3 | ×7 | ×3 | ×7 |

| 13 | 8 | 14 | 12 | 6 | 15 |
| ×7 | ×9 | ×9 | ×7 | ×3 | ×7 |

T7.4

Name: _____

$3\overline{)18}$ $3\overline{)30}$ $3\overline{)24}$ $5\overline{)60}$ $3\overline{)15}$

$9\overline{)27}$ $7\overline{)35}$ $3\overline{)21}$ $9\overline{)36}$ $7\overline{)21}$

$3\overline{)12}$ $5\overline{)20}$ $11\overline{)121}$ $7\overline{)84}$ $7\overline{)56}$

$3\overline{)39}$ $5\overline{)35}$ $11\overline{)132}$ $9\overline{)81}$ $3\overline{)27}$

Multiplication and Division © 2005 • www.zephyrpress.com

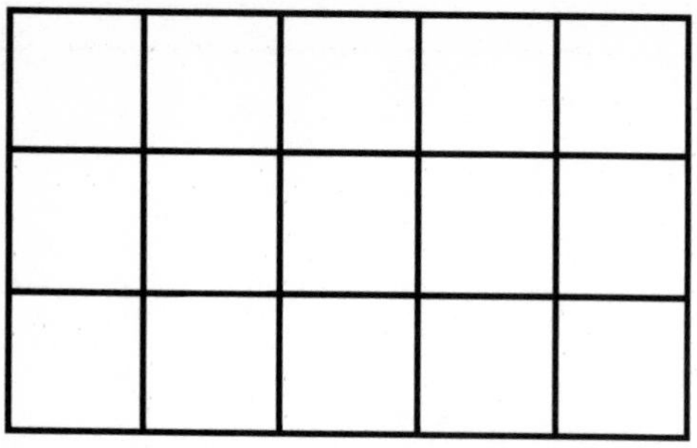

Name: ______________________

13 ×7	14 ×3	7 ×7	11 ×11	6 ×9	12 ×5
7 ×9	12 ×9	9 ×9	13 ×9	14 ×7	12 ×3
5 ×3	5 ×7	13 ×3	5 ×9	8 ×7	13 ×5
14 ×9	15 ×3	6 ×7	12 ×7	6 ×9	8 ×9

T7.5

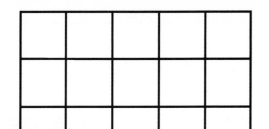

Name: _____

$9\overline{)27}$ $\qquad$ $3\overline{)24}$ $\qquad$ $9\overline{)36}$ $\qquad$ $11\overline{)121}$ $\qquad$ $7\overline{)56}$

$5\overline{)50}$ $\qquad$ $11\overline{)33}$ $\qquad$ $11\overline{)132}$ $\qquad$ $7\overline{)42}$ $\qquad$ $5\overline{)30}$

$5\overline{)40}$ $\qquad$ $9\overline{)81}$ $\qquad$ $5\overline{)45}$ $\qquad$ $9\overline{)72}$ $\qquad$ $7\overline{)35}$

$9\overline{)90}$ $\qquad$ $3\overline{)27}$ $\qquad$ $5\overline{)75}$ $\qquad$ $9\overline{)54}$ $\qquad$ $9\overline{)81}$

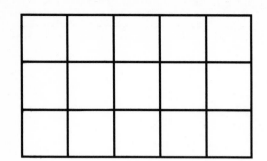

$9\overline{)27}$ $3\overline{)24}$ $9\overline{)36}$ $11\overline{)121}$ $7\overline{)56}$

$\begin{array}{r} 9 \\ \times 3 \\ \hline \end{array}$ $\begin{array}{r} 8 \\ \times 4 \\ \hline \end{array}$ $\begin{array}{r} 9 \\ \times 7 \\ \hline \end{array}$ $\begin{array}{r} 7 \\ \times 5 \\ \hline \end{array}$ $\begin{array}{r} 9 \\ \times 11 \\ \hline \end{array}$

$5\overline{)40}$ $9\overline{)81}$ $5\overline{)45}$ $9\overline{)72}$ $7\overline{)35}$

$\begin{array}{r} 9 \\ \times 5 \\ \hline \end{array}$ $\begin{array}{r} 7 \\ \times 3 \\ \hline \end{array}$ $\begin{array}{r} 7 \\ \times 6 \\ \hline \end{array}$ $\begin{array}{r} 9 \\ \times 8 \\ \hline \end{array}$ $\begin{array}{r} 8 \\ \times 3 \\ \hline \end{array}$

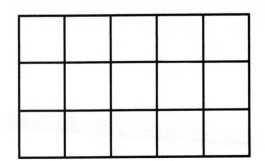

Name: _____

$$9 \times 4$$ $$8 \times 6$$ $$7 \times 4$$ $$6 \times 5$$ $$11 \times 7$$

$$8\overline{)72}$$ $$6\overline{)42}$$ $$4\overline{)56}$$ $$10\overline{)100}$$ $$8\overline{)64}$$

$$8 \times 5$$ $$12 \times 4$$ $$9 \times 6$$ $$9 \times 9$$ $$7 \times 7$$

$$6\overline{)48}$$ $$12\overline{)24}$$ $$4\overline{)24}$$ $$2\overline{)30}$$ $$12\overline{)144}$$

$2\overline{)14}$ $6\overline{)36}$ $12\overline{)156}$ $10\overline{)90}$ $4\overline{)52}$

$\begin{array}{r} 6 \\ \times 7 \\ \hline \end{array}$ $\begin{array}{r} 8 \\ \times 10 \\ \hline \end{array}$ $\begin{array}{r} 6 \\ \times 12 \\ \hline \end{array}$ $\begin{array}{r} 4 \\ \times 12 \\ \hline \end{array}$ $\begin{array}{r} 5 \\ \times 7 \\ \hline \end{array}$

$2\overline{)20}$ $6\overline{)84}$ $10\overline{)110}$ $8\overline{)96}$ $4\overline{)44}$

$\begin{array}{r} 12 \\ \times 6 \\ \hline \end{array}$ $\begin{array}{r} 7 \\ \times 12 \\ \hline \end{array}$ $\begin{array}{r} 7 \\ \times 8 \\ \hline \end{array}$ $\begin{array}{r} 9 \\ \times 12 \\ \hline \end{array}$ $\begin{array}{r} 8 \\ \times 9 \\ \hline \end{array}$

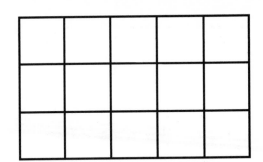

Name: _____

$9\overline{)90}$ $3\overline{)27}$ $5\overline{)75}$ $9\overline{)54}$ $9\overline{)81}$

$$\begin{array}{r} 5 \\ \times 12 \\ \hline \end{array}$$ $$\begin{array}{r} 12 \\ \times 7 \\ \hline \end{array}$$ $$\begin{array}{r} 8 \\ \times 2 \\ \hline \end{array}$$ $$\begin{array}{r} 7 \\ \times 9 \\ \hline \end{array}$$ $$\begin{array}{r} 8 \\ \times 7 \\ \hline \end{array}$$

$5\overline{)50}$ $11\overline{)33}$ $11\overline{)132}$ $7\overline{)42}$ $5\overline{)30}$

$$\begin{array}{r} 6 \\ \times 8 \\ \hline \end{array}$$ $$\begin{array}{r} 8 \\ \times 8 \\ \hline \end{array}$$ $$\begin{array}{r} 14 \\ \times 10 \\ \hline \end{array}$$ $$\begin{array}{r} 9 \\ \times 2 \\ \hline \end{array}$$ $$\begin{array}{r} 6 \\ \times 4 \\ \hline \end{array}$$

$$\begin{array}{r}12\\ \times5\\ \hline\end{array}\qquad \begin{array}{r}6\\ \times4\\ \hline\end{array}\qquad \begin{array}{r}7\\ \times9\\ \hline\end{array}\qquad \begin{array}{r}3\\ \times12\\ \hline\end{array}\qquad \begin{array}{r}6\\ \times9\\ \hline\end{array}\qquad \begin{array}{r}12\\ \times6\\ \hline\end{array}$$

$$9\overline{)27}\qquad 7\overline{)35}\qquad 3\overline{)21}\qquad 9\overline{)36}\qquad 7\overline{)21}$$

$$\begin{array}{r}6\\ \times3\\ \hline\end{array}\qquad \begin{array}{r}8\\ \times9\\ \hline\end{array}\qquad \begin{array}{r}11\\ \times7\\ \hline\end{array}\qquad \begin{array}{r}8\\ \times8\\ \hline\end{array}\qquad \begin{array}{r}9\\ \times5\\ \hline\end{array}\qquad \begin{array}{r}6\\ \times8\\ \hline\end{array}$$

$$9\overline{)27}\qquad 3\overline{)24}\qquad 9\overline{)36}\qquad 11\overline{)121}\qquad 7\overline{)56}$$

T7.8

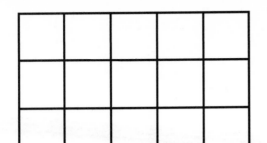

$3\overline{)39}$ $5\overline{)35}$ $11\overline{)132}$ $9\overline{)81}$ $3\overline{)27}$

$\begin{array}{r} 7 \\ \times 6 \\ \hline \end{array}$ $\begin{array}{r} 8 \\ \times 3 \\ \hline \end{array}$ $\begin{array}{r} 9 \\ \times 8 \\ \hline \end{array}$ $\begin{array}{r} 7 \\ \times 7 \\ \hline \end{array}$ $\begin{array}{r} 8 \\ \times 5 \\ \hline \end{array}$ $\begin{array}{r} 4 \\ \times 12 \\ \hline \end{array}$

$9\overline{)90}$ $3\overline{)27}$ $5\overline{)75}$ $9\overline{)54}$ $9\overline{)81}$

$\begin{array}{r} 5 \\ \times 12 \\ \hline \end{array}$ $\begin{array}{r} 9 \\ \times 6 \\ \hline \end{array}$ $\begin{array}{r} 6 \\ \times 7 \\ \hline \end{array}$ $\begin{array}{r} 8 \\ \times 6 \\ \hline \end{array}$ $\begin{array}{r} 9 \\ \times 4 \\ \hline \end{array}$ $\begin{array}{r} 7 \\ \times 12 \\ \hline \end{array}$

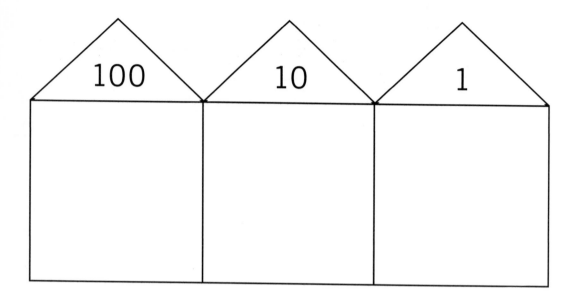

100   10   1

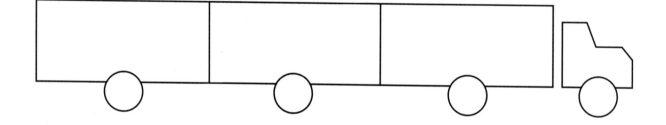

T8.1

Solve.

26 ×2	62 ×5	85 ×4	47 ×3	23 ×8
54 ×6	76 ×9	39 ×2	48 ×7	68 ×5
23 ×3	37 ×6	66 ×4	54 ×8	72 ×5
83 ×2	95 ×5	45 ×9	13 ×6	56 ×3

Solve.

78	43	42	27	65
×3	×9	×5	×8	×3

39	45	76	38	64
×6	×4	×2	×7	×4

85	47	56	24	48
×6	×8	×5	×9	×7

37	95	49	51	73
×5	×8	×9	×6	×7

Name: _____

Solve.

427	448	223	554	751
×3	×9	×5	×8	×3

853	248	793	127	692
×6	×4	×2	×7	×4

267	152	645	234	752
×6	×8	×5	×9	×7

4137	2395	1451	2453	1133
×5	×8	×9	×6	×7

T8.4

Multiplication and Division © 2005 • www.zephyrpress.com

Also available in the *Kid-Friendly Computation* series

Addition and Subtraction   Place Value

By Sarah Morgan Major

Kids can say good-bye to math anxiety with the help of the learning techniques introduced in this series of teacher aides. Drawing on visual and kinesthetic learning strategies instead of abstract concepts, even students who have always struggled with computation can become math whizzes. Each book is designed to replace the drudgery normally associated with rote memorization with the satisfaction and success gained from engaged, hands-on learning. The key is recognizing visual patterns that exist between numbers when they are arranged in global "wholes." These visual patterns provide multiple pathways for comprehension and recall, and are especially suited for kids who have had little success with traditional teaching techniques. *Addition and Subtraction* begins with number recognition and counting, and continues through single-digit computation, number patterns, number sense, and number families. *Place Value* takes basic addition and subtraction into multi-digit computation, and includes strategies for word problems.

Ages 7–10, 8½ × 11
60 line drawings (in each)
Paper, $29.95 (CAN $41.95) each

156976199X, 192 pages (*Addition and Subtraction*)
1569762007, 224 pages (*Place Value*)

Available at your local bookstore or order by calling (800) **232-2198**, or online at **www.zephyrpress.com**

Zephyr Press

Distributed by Independent Publishers Group
www.ipgbook.com